BARE NECESSITIES

Sahar Mansoor is the founder and CEO of Bare Necessities, a zero-waste social enterprise. She is a circular-economy nerd, a voracious learner and a curious traveller.

Sahar has obtained accredited degrees on environmental topics in India, the USA and England, becoming a University of Cambridge alumni.

Her business offers zero-waste personal care, lifestyle, home care and educational products and services, including online sustainability courses called 'Zero Waste in 30' and 'Sustainability in 30'. She has also volunteered in Guatemala, Jamaica and the Democratic Republic of Congo for humanitarian projects. Her time working at the World Health Organization in Geneva and SELCO Foundation, on implementing decentralized renewable energy projects in rural Karnataka, enabled her to build Bare Necessities from the ground up. Her work has been recognized by Google India, Al Jazeera, NDTV, *Vogue*, *Elle*, *Femina*, *India Today*, among others.

Sahar has a 500-gram jar of waste that she brings to the sustainability workshops she conducts, which is all the waste she has produced in the last five years. For more, visit https://barenecessities.in/.

Tim de Ridder began his environmental career in Australia, prior to moving to South East Asia and South Asia. He has a master's degree in environmental management and sustainability from Monash University, Australia. He has organized a wide range of workshops on sustainability, with a focus on the circular economy, the creation of an online education course on waste in India and community-engagement initiatives in Cambodia and Malaysia.

He has also written widely on environmental topics, including a research paper on the livelihoods of low-income communities and the benefits of carbon farming for indigenous Australians.

Bare Necessities is the second book he has authored after his self-published *Tales from the Roadside*, a memoir about a solo bicycle trip across Europe that connected him to the world in profound ways.

BARE
NECESSITIES
how to live a
Zero-Waste Life

Sahar Mansoor
&
Tim de Ridder

PENGUIN BOOKS

An imprint of Penguin Random House

PENGUIN BOOKS

USA | Canada | UK | Ireland | Australia
New Zealand | India | South Africa | China | Singapore

Penguin Books is part of the Penguin Random House group of companies
whose addresses can be found at global.penguinrandomhouse.com

Published by Penguin Random House India Pvt. Ltd
4th Floor, Capital Tower 1, MG Road,
Gurugram 122 002, Haryana, India

Penguin
Random House
India

First published in Penguin Books by Penguin Random House India 2021

Illustrations by Mouli Paul

10 9 8 7 6 5 4 3 2

ISBN 9780143451174

Typeset in Bembo Std by Manipal Digital Systems, Manipal
Printed at Manipal Technologies Limited, India

www.penguin.co.in

MIX
Paper | Supporting
responsible forestry
FSC® C043100

CONTENTS

INTRODUCTION

'Hi, my name is Sahar.

I will be your guide through this book. I will be sharing many stories that I have learnt, which have helped me achieve a zero-waste lifestyle. "You wrote a book?" you may be asking! Well, yes. You see, I'm so passionate about this topic that I have even created a zero-waste social enterprise to make mindful and low-impact living accessible to more people.

I have always wanted to create a hub of information for people to use, and I hope this guidebook does just that. Throughout the nine chapters there are more than eighty tips and tricks that can help you move towards a zero-waste lifestyle. Added to this, in the middle section of the chapters, are over two dozen recipes and suggestions of resources that you could use to reduce waste in your life. There is also the 'Zero-Waste Library', which is one of my favourite sections. I've had a little fun with the names and creative ideas for you. There is a broad range of things here, a library of over fifty resources! Pretty useful, right?

So, why me?

That's a big question! I'll tell you a little about it here and then much more in each chapter. I'm a passionate environmentalist who has learnt many things (from many failed experiments!) about reducing waste that are useful to share. Knowledge sharing is key when it comes to helping reduce waste all over the world. So, let's begin.

When I was young, I loved being outdoors, exploring nature. I really think a lot of my passion and environmental drive took shape back then. I also had a lot of luck having had the opportunity to learn from the finest environmental researchers and professors at Loyola Marymount University!

They exposed me to environmental and social justice issues that I cared about more and more each time I read about them. In one of my classes I learnt about a woman named Bea Johnson who was living a zero-waste life. It was inspiring, but at that time I couldn't believe that I too could achieve what she had done. Certainly not while trying to keep my grades up and working three jobs!

Over the years I followed my passion for the environment to another degree, this time based in the United Kingdom. I found that many of the lessons I was learning, from the inspiration of Bea Johnson to life lessons from my fellow students, came back to how, at least in the mainstream, there were only a few Indians speaking about reducing waste in the country. I was still inspired by the knowledge I was gaining, of course, but I didn't know exactly how it would relate when I returned home to India.

I couldn't see how the home-delivery, personal-care or kitchen waste I witnessed in California, at college or when I visited my big sister (who lived there with her family, including my little niece), would connect to my home in Bengaluru. I was at a loss, to be honest.

I still was when I returned home to Bengaluru, a short while after completing my second degree. Bengaluru was my home town, where I had grown up with my mum, two older sisters and (in my opinion) the best cat in the world. I had all the lessons about the environment under my belt, but I was not sure what they meant, practically speaking, in India.

My first job after my return was with SELCO Foundation, an organization which focuses on providing energy-access solutions to various underserved communities (including migrants).

I worked all across the state of Karnataka, meeting people, learning from them and exposing myself to new areas of my home state that I had no idea about while growing up, such as the waste-picking communities.

All of this was amazing to me. Access to electricity and social justice issues that related to the waste-picking communities were so important to me. Yet, over the two years that I worked with SELCO Foundation, I found my way back to focus on reducing waste. Some of this came to me while assessing the amount of waste I was generating after I visited the waste-picking communities, where the workers sorted through piles of discards with their bare hands! I learnt quite a bit through reminders of what people were doing to reduce waste in other parts in the world, such as zero-waste champion Lauren Singer, who was based in New York, USA. She was the same age as me (twenty-five at that point) and had created a zero-waste business. Talk about inspiring!

I didn't plan to create an enterprise like Lauren. Instead, I started a step-by-step process that began by learning zero-waste recipes, such as making soap and toothpaste. From that point, I followed my ideas to show people how to create their own things at workshops and markets. Gradually, I expanded my product range and met more people in the industry.

Since 2015, the sustainability space has expanded in India. I moved on from SELCO and set up my own business in 2016. I am happy to say that knowledge and resources are far more easily available these days. Sustainability is becoming commonplace in mainstream media, which is really encouraging. This means that more people want to get involved and make a difference.

So, why me?

Well, I wouldn't say that I am unique in the global zero-waste space (and that's really a good thing; the more the number of people involved, the more likely we are to make a difference), but I am in a position where I can share my story with you through this guidebook. I can now use the lessons from my journey to provide you with something that I did not have: an Indian perspective on living a life that values the planet and all the people, plants and animals we share it with.

My hope is that I'll be there as a helping hand, if you need one, as you start off on your very own journey.'

< IX >

HOW TO USE THIS GUIDEBOOK

Choosing to open this guidebook is the first step of your journey towards a more sustainable lifestyle. You will be guided along by insights from a zero-waste expert, your guide, Sahar.

But first, let's walk you through the structure of the book.

Each chapter is designed to flow outwards from the areas that you encounter immediately upon waking each day, namely your bathroom (or personal-care areas), your closet (or fashion choices) and your kitchen, which can include a variety of waste from organic to inorganic materials. The next step in the process is to look at broader areas in your home and at the wasteful habits that may be a part of it, such as your gifting or festival practices. The broadest of the areas is 'community', which involves many things that you encounter on a day-to-day basis outside of your home. Additionally, Chapters 7 to 9 include discussions on broad topics. Lastly, you have access to a reading list at the end of the book.

The guidebook's process is designed to encourage you to think about waste-producing areas in stages, in interactive ways. Sahar will share lessons and provide insights based on her own experience.

Hopefully you will find the conversations, activities and information absorbing, whether you are based in India or outside. As mentioned before, knowledge sharing is important and fun for everyone. Sahar has certainly enjoyed sharing part of her story. With a little luck, enough of it will resonate with you, which in turn will encourage you to share your own thoughts and ideas with the people around you.

Detailed below is a brief description of the chapter structure:

The Perfect Life . . .?

The first section of each chapter details a hypothetical situation. This section is to encourage you to reflect on current situations by thinking about examples that produce waste in each of the eight areas.

What Is in Your Waste?

The first activity in the chapter is to test your knowledge using information that you currently have. Try it out and see how it goes. There are no wrong answers. You can always come back and update areas as you learn more about certain situations that create waste.

This subsection is the first time that you will see an activity sheet in each chapter. Attempt to assess the waste you generate in the focus area before reading on. Once you have gained more knowledge from Sahar, the recipe section, the tips and tricks component, or as part of the 'Zero-Waste Library', you can revisit it.

What Resources Are Available?

In the third subsection of each chapter, Sahar will share knowledge about the topic in order for you to undertake Activity #2. This activity is designed to promote thinking about everything as 'resources' instead of 'waste'. Similarly, you can learn more throughout the chapter and return with new knowledge.

How to Move Towards a More Sustainable Lifestyle?

The fourth subsection of each chapter provides you with a range of knowledge, either as part of Sahar's guidance and personal stories and/or as zero-waste recipes, or ideas, events, activities, people and/or organizations to learn from in order to move towards a zero-waste lifestyle.

Additionally, a component named 'Zero-Waste Tips and Tricks' and two illustrated representations of consumption and production systems are also provided. Namely, this is a waste-generating model, called the linear economy, and a resource-saving model, called the circular economy. The pictures follow a process from chapter to chapter to help you learn about how things are made, used and discarded using the two methods. They also show the impact of choices relating to these areas at each stage, which relate to information in the 'Why Is It Harmful?' section below.

The combination of resources in this section is designed to help you in the activities peppered throughout the guidebook and to provide you with the tools that you could use to transition to more sustainable practices.

Why Is It Harmful?

This is the most difficult section in each chapter because it discusses both broad and small topics about waste. It is designed to be challenging and to teach you about a range of topics that relate to the chapter.

The information in this section comes from various resources that relate to specific areas of the world and also highlight how interconnected every person on the planet is. You can use the information in various ways, be it on the activity pages or simply as discussion points with your friends and family.

To help you out in this section, Sahar will provide an overview. Each section has been segmented into a discussion piece that relates to the topic and the overall theme of the chapter. A lot of the information you will learn in this section, and others, can be used in other chapters too.

What Can You See in Your Waste Now?

The third activity is located in this subsection where Sahar will briefly join you for the last time in each chapter. She will reflect on some of the lessons that you have learnt until that point and provide an overview of the final activity. Try and challenge yourself to get the answers right and/or undertake more research in the guidebook or in other sources if you need more assistance. Most of the answers are available in the pages of the book, such as those recommended by Sahar or in the 'Zero-Waste Library' section.

Zero-Waste Library

The final section varies from chapter to chapter based on the resources that you need for the activity sheets. There are recipes, activities and many more pieces of information that you can learn from. Start with a trial at home and eventually implement it in your life.

Enjoy the process and have fun with all of the resources that you have been provided with!

*

As noted above, you have taken the first step by opening the guidebook. By the time you reach the final pages, you will truly be at the start of your next big adventure.

> 'Stories have always been important to me. I hope that one day I will be able to hear your story about how you live a zero-waste lifestyle, just like you will hear mine.'

use a neem comb or hairbrush

use a safety razor instead of a disposable one

make your own toothpaste

ZERO-WASTE PERSONAL CARE

stop using cotton earbuds and wet wipes. Use reusable wipes instead

use a compostable toothbrush

use plastic-free feminine hygiene products

use personal care products that come in eco-friendly packaging

use cotton napkins instead of tissues

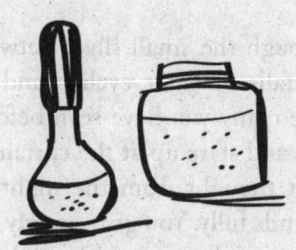

CHAPTER 1

PERSONAL CARE

'Though one wrapper of soap or one bottle of shampoo might not seem to matter, it can definitely make a difference when multiplied by over a billion people who live in India and who comprise one-seventh of the world's population.'[1]

The Perfect Life . . .?

Light seeps in through the small sliver between the curtains. The single strand falls on your eyelids and you stir. It is a morning much like many you have seen before. You stretch a little, rub your eyes and stare up at the curtain. For a moment, you recall your last thought from the night before that you should close the blinds fully. You grin smugly at your situation. A few more minutes of sleep would have done you well, you think, but it wasn't to be. Instead you move your legs beneath the sheets, place your feet on the ground and allow the rest of your body to follow.

Moments later you are standing in front of the mirror in your bathroom with the soft gurgle of your toilet fading away. You wash your hands with soap and splash water over your face before moving to the shower.

The water rushes all around, the soap and shampoo bubbles flow to the ground, dancing around your naked feet before edging past and pursuing the downward trend into the drain. Then, in a practised motion that you have perfected over time, you shave away all the unwanted hair and drop the blades into the bin.

After your shower, you stand in front of the mirror and fix your hair with more products. Next, you pat, dab and rub what you need to on your skin. You look at yourself and all of the items that you have on your counter. You smile. The sun seeping in from between the gap in the curtains has put you in a good mood. It is going to be a glorious day, you think.

Then you take one final look at your countertop and you know that every item you hold dear as part of your morning routine is there. Some in use, some empty, some brand new. The ones in use are at the front. The empty ones are piling up in the bin with the blades that you casually threw in a few moments before. The new ones are stacked in neat rows. There are plenty of them since they were available at a special price, for an excellent deal. You smile once more and shut the bathroom door behind you.

What Is in Your Waste?

'Packaging wastes are a very visible part of environmental problems . . . The package is not noticed during purchase, transport, and use of the product—in fact, it is not noticed until the minute the product is consumed and the package has fulfilled its function and turns into waste. At that minute, the package is already seen as an environmental burden, wasting resources.'[2]

'If you come and look at my bathroom now, you may think that it was a simple process to make it waste-free. I can assure you it wasn't . . . I experimented over and over again. My mum's kitchen was my first laboratory. A quick thanks to her for being so patient through all the failed, messy experiments during my early days. Our bathrooms are one place where we have full control over what we do and how we do it. It is the first place that we run to each morning and the last place we visit before going to bed. Change there can help lay the groundwork for a lot more in life.

I had this crazy realization at one point, when researching about all the waste generated by the products we generally use in our bathrooms, like toothpaste tubes and bottles of shampoo. In addition to these products' packaging being constructed with multi-layered plastic (practically impossible to recycle), what is to be noted is that our skin, which is the largest organ,[3] absorbs over 60 per cent of what we expose it to or put on it.[4] This was honestly an eye-opener. It was scary to think about the products I was using.

I read and researched about all the ingredients and learnt how harmful the chemicals were to our environment. Both the chemicals and the plastic containers that I was using floated through my racing mind until they fell to the bottom of the sea. I knew that I had to make a change, I couldn't have that knowledge and not do anything about it.'

NOW IT'S YOUR TURN TO ASSESS!

Try out Activity #1 with your current knowledge. View it as an initial way to see how much you know. Once you have learnt more through this chapter, you can return to it in order to see how much your knowledge has developed.

ASSESS YOUR
PERSONAL CARE WASTE:
ACTIVITY #1

Follow the example sheet below by choosing a few
of your own products to assess.

	Example A	Example B	Example C
Is the container recyclable or reusable?			
Is the product organic?			
Does the product contain harmful chemicals that can harm you and/or the environment?			
Do you know all the ingredients that are in your product?			

- If yes: share the solutions with friends and family.
- If no: research options.

Use the ideas from this sample to assess your waste. You can draw your own sheet based on
this and create other questions for this assessment based on your needs.

What Resources Are Available?

'After assessing the waste produced, I learnt to make my own home-made toothpaste (the recipe is in the section below) and realized that I needed to try things out to learn about the resources we have in our country. There were many products I tried, which I learnt had previously been used in India, such as the *miswak* stick, the first toothbrush in the history of humankind.[5] Interestingly, it was used by the ancient civilizations—from the Babylonians to the Greek to the Romans and the ancient Egyptians.[6]

I'm sure you're wondering what the benefits of a simple miswak stick can be, so here they are:

- Besides being 100 per cent organic and completely compostable, the miswak stick is full of natural minerals such as potassium, sodium, chloride, sodium bicarbonate and calcium oxides. These ingredients are said to strengthen the tooth enamel.
- The bark contains an antibiotic which suppresses the growth of bacteria and the formation of plaque.
- Research shows that regular use of miswak significantly reduces plaque, gingivitis and growth of cariogenic bacteria.
- It naturally strengthens and protects the enamel with resins and mild abrasives for whiter teeth and fresher breath.
- Miswak sticks massage the gums, giving you healthier teeth.
- For all coffee and tea lovers, miswak reduces caffeine stains too!
- Compared to the standard toothbrush, the form of the twig makes it easier to get to the hard-to-reach places.

Pretty cool, isn't it? Yet, as sustainable and accessible as it was, I honestly missed the sensation of brushing. I asked my sister, who lived in Copenhagen, to get me a bamboo toothbrush because they weren't available in Bengaluru. I had to wait four long months before she got here with my request. Thank you, Sis!

In 2015, I started using it and have never looked back. Fast forward a few years, bamboo toothbrushes are readily accessible in India.

I began looking at all my products differently when I found that my toothbrush could be composted to produce quality soil that would help grow plants. This, to me, when compared to the impact single-use products have on the environment, was startling.

There were many other products I came across, where I simply had to ask questions about the production of the material and the packaging. This became highly important to me, but knowing what happened after I used these products became even more important. I kept visualizing the waste pickers or the fish in the ocean having to deal with my discards. What type of person would I be if my waste was a part of this problem? I began to think.

The first steps were hard, but there were a lot of resources available all around, for example, products made from coconut oil (easily found throughout south India) to lavender (from Kashmir). During this stage of my journey, I began to identify with my roots profoundly. India is such a phenomenal land for natural ingredients that can help keep our bodies clean without contributing harmful materials to our environment.

The evaluation of my waste, research and experimentation (especially learning from failed attempts) paid off. I'd really encourage you to do the same while you think about the information you're learning now.'

How to Move Towards a More Sustainable Lifestyle?

'It was all a step-by-step process, starting with evaluating and collecting the waste I was generating in a cotton bag with the products that I used and finished (it didn't take long to fill the bag). I was shocked at the pace at which it was filling up once I started paying attention! I spoke to my Nani (maternal grandmother) about the products she had used before shampoo started being sold in plastic bottles, and I attended a soap-making workshop to learn more about what was available close to my home. This new knowledge led me to undertake really messy experiments in my mum's kitchen (oh, the looks she gave me!). I then set realistic goals to find and use resources sourced from local locations, which would not have adverse effects on the environment. One by one, my experimentation led to results that were good for my body and for the environment. A couple of my favourites are detailed below. Try them out. I hope you enjoy them.'

TRY THIS ACTIVITY OUT!

In this chapter there are a number of suggestions that you can use for Activity #2. You can read ahead and return with new knowledge or assess yourself now and return to this exercise later.

< 8 >

ASSESS YOUR
PERSONAL CARE RESOURCES:
ACTIVITY #2

Find one to three products that you can make yourself. Use the suggestions in the guide book or research to find more. Fill in each section of this activity to record your achievements.

The resources that I have learnt about are:

 Peppermint Party Toothpaste

 Lemongrass Bath Salts

 Dessert Dry Shampoo

Draw a picture from your activity to show friends and family:

The resources needed for this activity are:

 Baking soda, coconut oil, peppermint oil.

 Epsom salt, coconut oil, lemongrass oil, rose petals.

 Cornflour, cocoa powder.

Record where you learnt about these resources:

 The guide book's 'Zero-Waste Recipes' section.

Record who you shared your success with:

 My family! They have made their own, too, with my help.

Use the above ideas to assess your resources and then create your own activity sheet.

ZERO—WASTE RECIPES

Peppermint Party Toothpaste[7,8]

You will need:

- 1 part baking soda,
- 1 part coconut oil,
- A few drops of peppermint essential oil.

All you need to do to get naturally clean teeth is to mix it all together in a bowl until it becomes a paste and then place it in a reusable container for the next time you brush your teeth.

While this may taste and appear to be different from the toothpaste you're used to, it is worth trying out. There are other zero-waste options available if this taste is not for you. You can always research online to learn more.

To use the Peppermint Party Toothpaste:

- Scoop a tiny dollop with a teaspoon on to your toothbrush (you could start using this and a bamboo toothbrush at the same time) before every use.

Lemongrass Bath Salt

You will need:

- 1 part rock salt/Epsom salt/pink Himalayan salt (or a combination of all),
- A small drizzle of coconut oil,
- A few drops of lemongrass essential oil,
- Rose petals.

For a naturally rejuvenating bath, follow this method:

- Gradually mix each of these ingredients in a bowl in chronological order,
- Place it into a reusable container for use when you bathe next.

Dessert Dry Shampoo

You will need:

- 1 part cornflour,
- 1 part cocoa powder,

All you need to do is put the cocoa powder into a bowl, followed by the cornflour.

This one is really easy to make, with no complications involved:

- Stir the two ingredients with a spoon until there are no distinct ingredients, that is, you can no longer see any white from the cornflour. The mix should be light brown in colour.
- Place this in a container for you to use when you need it.[9]

'There are other great ways to take positive steps forward, including visually seeing the impacts of our choices and learning why they're harmful to the planet. On the next two pages are two illustrated overviews, one of a wasteful system and the other of a resource-saving system. In the following chapters, we'll get a chance to view each step of the process individually.'

ZERO—WASTE TIPS AND TRICKS

- Use a wooden comb or hair brush,
- Use a safety razor instead of a disposable one,
- Make your own toothpaste,
- Use a compostable toothbrush, e.g., a bamboo brush,
- Stop using cotton ear buds,
- Make products from scratch, e.g., make-up remover,
- Use plastic-free feminine hygiene products,
- Stop using personal-care products that come in plastic. Instead, find alternatives that use glass, paper, etc.

PERSONAL CARE-LINEAR SYSTEM

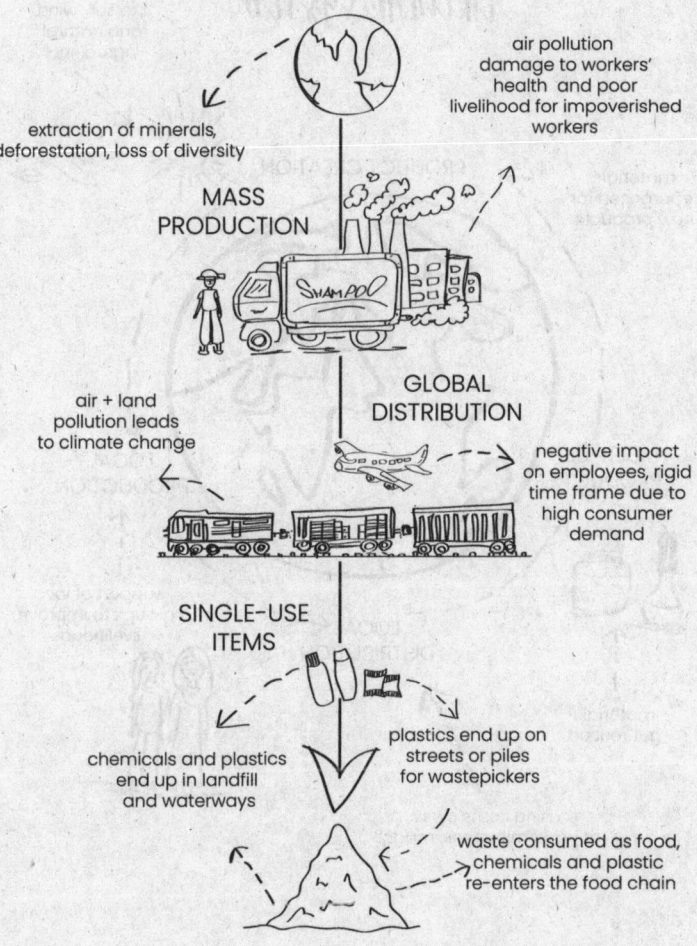

extraction of minerals, deforestation, loss of diversity

air pollution damage to workers' health and poor livelihood for impoverished workers

MASS PRODUCTION

SHAMPOO

GLOBAL DISTRIBUTION

air + land pollution leads to climate change

negative impact on employees, rigid time frame due to high consumer demand

SINGLE-USE ITEMS

chemicals and plastics end up in landfill and waterways

plastics end up on streets or piles for wastepickers

waste consumed as food, chemicals and plastic re-enters the food chain

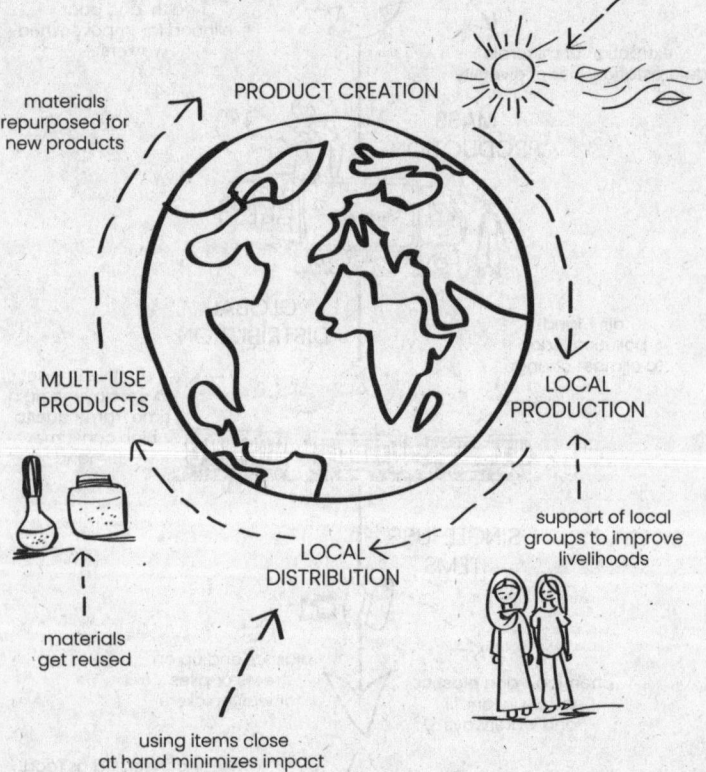

PERSONAL CARE - CIRCULAR SYSTEM

natural sources like sun, wind and natural ingredients

PRODUCT CREATION

materials repurposed for new products

MULTI-USE PRODUCTS

LOCAL PRODUCTION

materials get reused

LOCAL DISTRIBUTION

support of local groups to improve livelihoods

using items close at hand minimizes impact

Why Is It Harmful?

'I hope you've learnt about some really unique things to do to reduce waste in your bathroom so far.

The next stage that I find really useful is to understand why things like the ingredients in personal-care items or everyday items, like plastic,[10] are really harmful. We'll be covering a number of topics in the coming chapters in this regard. Learning about the reasons why things are harmful, or not, will allow you to make informed choices.'

There are, most likely, many different products that you use on a regular basis as part of your personal care. These may range from cosmetics to moisturizers to soaps to toothpastes. Then there is a toothbrush, razors, and if you are a woman, feminine hygiene products.[11, 12]

Of course, you may have more or less depending on how you choose to live. So, for now, it will be best to keep it in two major categories: those that are applied to your body (such as soaps and toothpaste) and those that are tools (such as a razor or toothbrush).

Ingredients

'There is some evidence that paraben (a chemical used in cosmetics) is partially at fault for killing off coral, and more than a few scientists believe that this chemical is a hormone disruptor in dolphins and other marine wildlife.'[13]

Starting off with the products that you usually lather on to your skin, think for a moment about what ingredients go into making them. Palm oil[14] is a ubiquitous ingredient found in a vast array of personal-care products such as hair conditioners (and many other things outside of your bathroom, i.e. Nutella).[15] This ingredient, from a tree that was originally transported from western Africa to Indonesia and Malaysia where it now grows abundantly in crop farms,[16] has a detrimental impact on natural forests, the indigenous people that live there and the flora and fauna of the locations it is found in.[17] Notably, the impacts are not only found in the local vicinity, but also globally. How does your decision to use a product containing palm oil matter? For an oil extracted from a palm tree, it relates to you as a consumer and how you view extended producer responsibility,[18] expanded below in relation to cosmetics. Simply put, are you a person who will be a catalyst for change on a global scale because of your choices at an individual level or not?

Perhaps it is difficult to relate to deforestation at a location that you have never seen. But think about your bath soap, for instance. Have you been using it for some time? Does it have an ingredient that you do not know of?

An international study has highlighted that some soaps contain a chemical that is an endocrine disrupter, which has the potential to alter hormonal activity, which in turn has been reported to cause numerous problems including cancer, reproductive failure and developmental anomalies.[19] Does this seem more personal? Understanding the ingredients[20] is the first step. From there you can urge, through your demand, manufacturers to be more environmentally responsible for all of the ingredients that they use.[21]

Substantial work by scientific studies, regulatory bodies and businesses to understand the effects of personal-care products on people and the environment[22] has been undertaken in the past decade, as awareness of current practices relating to the ingredients used in products has grown. In addition, conscious consumers in recent years have not only created the need for new businesses that produce environmentally safe

< 16 >

products but also the surge in demand has led to businesses refusing to change to become obsolete.[23]

Many of these issues, before active citizens came into the picture and began promoting transparency and change, come down to businesses performing 'business as usual' practices. The cosmetic industry is a perfect example. When cosmetics began to be used more widely, post-World War II,[24] large businesses focused on better living standards through the use of newly discovered chemicals. A 'better life' through chemical innovation, so to say. The major issue was that the chemicals were not rigorously tested to see what impacts they could have. This has led to their use becoming so widespread that it is almost as though the people who should have checked this situation dropped the ball on it.[25]

Today, when some regulatory policies are more stringent,[26] the manufacturers of these products strongly suggest that the quantities of the chemicals are small enough to not have a negative impact. This is despite the fact that many of the ingredients are known to be harmful, or even have unknown effects.

This situation also does not take into account the fact that consumers often use multiple products per day, which then increases the quantity of chemicals their bodies are exposed to. Nor does it consider the mix of chemicals from different products. Toxins mixed into a product during the manufacturing process end up polluting the environment when the products are used (described as the principle of 'toxins in, toxins out' by Annie Leonard in *The Story of Stuff*). This impacts both the environment and the person using the product,[27] yet manufacturers continue to market products as items that do no harm, in part because they are not held accountable for their greed.

This is a systemic issue that does not view the impacts of a product's lifecycle.[28] It is a situation that is hurting consumers all over the world. It needs to be addressed through awareness and action in the form of petitioning for transparency and responsibility[29] from those in charge of making the decisions to place certain products on the shelves or not.

Plastic

'Every day, India generates plastic waste that weighs as much as 150 large blue whales—the biggest animal known to exist.'[30]

The plastic found on or in products that you may use on a daily basis is alarming. The use of toothbrushes, for instance, has increased since the early last century when criteria introduced by the military during war time saw an increased use of the dental hygiene tool.[31] Originally, organic methods[32] were used in manufacturing. These manufacturing processes have transitioned to plastic being a key component, due to the cheap material and functionality of the polymer.[33] Advocates for change in this area point out that toothbrushes take over 400 years to decompose[34] and are rarely recycled, which means that they end up in landfills and pollute the environment.

Pollution caused by landfills that are over capacity from fast-moving consumer goods (FMCG), such as the toothbrush and other post-consumer products,[35] is regularly seen in rivers and oceans[36] all over the planet. Conversations about banning plastic, such as those in India during the latter half of 2019,[37] and an increased focus on this issue globally[38] will play an important step in stemming the plastic tide.[39]

India consumes 'an estimated 16.5 million tonnes'[40] of plastic, with approximately '80 per cent of the total . . . discarded immediately', which in turn means that waste finds a 'way to landfills, drains, rivers and . . . the sea.'[41] The heavy reliance on plastic is highlighted by the fact that the FMCG sector is the 'fourth-largest sector in the Indian economy with household

< 18 >

and personal care accounting for 50 per cent of FMCG sales in India'.[42] The increase has notably been 'led by a combination of increasing incomes and higher aspirational levels (with an) increased demand for branded products'[43] and an inefficient waste system.[44]

Transitioning to more sustainable methods from the current ones can be difficult to implement in the initial stages due to the influence of the stakeholders who have vested interests in the current system.[45] Clear roadmaps need to be in place to transition all countries away from the use of plastic and other harmful materials[46, 47] that are increasingly being used for the last half a century. Almost half of all plastic produced has been created since the year 2000, with 8.3 to 9 billion tonnes created since the 1950s, which is equivalent to more than four Mount Everests.[48]

From soaps to cosmetics to toothbrushes and many other consumables designed for one-time use-and-throw, the resulting effect[49] will often be that of microplastics polluting the environment.

Microplastics have been found to damage the entire food chain because the small fish at the bottom of the hierarchy see the tiny fragments (plastic bits smaller than 5mm)[50] as food.[51] Then, as the small fish are eaten, the plastic is consumed by predators all the way up the food chain, building into larger quantities at each step.

Therefore, two of the major pollutants from personal-care products: the harmful ingredients and plastic, impact the planet in the long term and your body almost immediately. The chemical residue from the products that you put on or in your body may end up in waterways that hold the water that communities drink, bathe their children in or utilize in order to grow vegetables to sell to you.[52] On the other hand, plastic contaminates the environment in numerous ways, from animals eating it to it degrading into microplastics that are found throughout the world.

What Can You See In Your Waste Now?

'This is all super intimidating, I know, but if we don't know why things are harmful, we may never initiate a change.

Try using this new knowledge for the final section in this chapter. If you're like me, you'll be able to assess things better in the second assessment than the first time.

Good luck! Remember, there are no wrong answers. It's all about trying your best, especially if you have just started. It's all about the process.

I'll meet you again in the next chapter with a few stories and ideas about the fashion industry.'

LOOK A LITTLE DEEPER
WITH THIS ACTIVITY!

Activity #3 is a four-part process, on separate question sheets, that builds on everything you have learnt throughout this chapter. The importance of this exercise is to understand and become aware of the processes that can help you transition to a sustainable lifestyle.

< 20 >

ASSESS YOUR
PERSONAL CARE WASTE:

What type of environmental impact do your products/services have?
Join the product/service to the problem using an arrow:

Product/Service:

Spray deodorant

Toothpaste

Soap containing
microbeads

Problem:

Damage to ground
water table

Air pollution

Plastic pollution

Use the ideas from this sample to assess your waste. The product/service
may relate to more than one problem. Fill in the lines below with your
products/services and environmental problems.

_____   _____

_____   _____

_____   _____

_____   _____

_____   _____

When thinking about the environmental impact, it is important to think of the number of areas that it could affect. Start on a small scale and work your way out. First, think about what it means to the micro-environment around you, then think larger and larger until you look at it from a global perspective. It will be beneficial to undertake some research online or in a library.

ASSESS YOUR
PERSONAL CARE WASTE:
ACTIVITY #3 QUESTION 2 OF 4

What system issues prevent change?
Join the product/service to the issue using an arrow:

Product/Service:	**Issue:**
Spray deodorant	No other products available
Toothpaste	No environmental policy in place
Soap containing microbeads	No affordable alternatives

Use the ideas from this sample to assess your waste. The product/service may relate to more than one issue. Fill in the lines below with your products/services and environmental issues.

_____   _____

_____   _____

_____   _____

_____   _____

_____   _____

A system is anything associated with the product or service that is interconnected with it, for example, the manufacturing unit that produces the product or the government office that provides the service. To learn more, conduct research online or in a library.

ASSESS YOUR
PERSONAL CARE WASTE:

What sustainable options are there to replace it?
Join the product/service to the solution using an arrow:

Product/Service:	**Solution:**
Spray deodorant	Natural, hand-made alternative
Toothpaste	Chemical-free items
Soap containing microbeads	Make your own product

Use the ideas from this sample to assess your waste. The
product/service may relate to more than one solution. Fill in the lines
below with your products/services and solutions.

_____   _____

_____   _____

_____   _____

_____   _____

_____   _____

A sustainable option is a product or service that will last longer and/or produce less waste. Think about
options such as products made from earth-friendly materials, or those that reduce waste through a supply
chain. To learn more, research online or in a library.

©Bare Necessities Zero Waste Solutions Pvt Ltd

How will you start using the sustainable option?
Join the product/service to the action using an arrow:

Product/Service:

Spray deodorant

Toothpaste

Soap containing
microbeads

Action:

Buy natural
products

Only purchase
chemical-free
products

Learn DIY recipes

Use the ideas from this sample to assess your waste. The
product/service may relate to more than one action. Fill in the lines below
with your products/services and actions.

_____   _____

_____   _____

_____   _____

_____   _____

_____   _____

This last step is all up to you. Make your choice, know the benefits and live a zero-waste lifestyle. You are more
likely to succeed with support from your friends and family.

ZERO—WASTE LIBRARY

Pipe Cleanse
Unclog Your Drain

You will need:

- ½ cup baking soda,
- ½ cup white vinegar,
- A wet cloth,
- Hot water.

Unclog your drain by:

- Pouring the baking soda down the drain,
- Follow with white vinegar,
- Cover the opening with the wet cloth,
- Let it sit for 5–20 minutes (based on how clogged your drain is),
- Pour some hot water down the drain.

This trick will work wonders for mild clogs!

Don't Stress about Stings
Home Remedy for Insect Stings

You will need:

- Epsom salts (no strict measurement),
- Water.

Ease the pain and swelling of an insect sting by:

- Mixing some Epsom salts in water,
- Soaking the affected part of your body in this mix for 10–15 minutes.

This is very effective for reducing pain and swelling, and helps you relax too!

Nails That Shine
Nail Cleanse

You will need:

- 1 teaspoon salt,
- 1 teaspoon lime juice,
- 1 teaspoon baking soda.

Whether you need it after gardening or cooking with turmeric, or maybe even after writing with an ink pen, or simply to refresh your nails in between painting cycles, follow this method:

- Mix the ingredients into a small bowl,
- Dip a small cotton cloth into the mix,
- Buff and clean your nails with the cloth (if you need more moisture after a nail or two, repeat the step above).

Cutie-cles
For Healthy Cuticles

You will need:

- 1 teaspoon honey,
- 1 teaspoon apple cider vinegar,
- 1 teaspoon coconut oil.

The honey moisturizes, the coconut conditions and the vinegar brings it all together in a truly effective method. All you need to do is:

- Combine the ingredients in a bowl,
- Using your fingers, rub this mixture around your cuticles,
- Let it sit for 10 minutes,
- Rinse with warm water.

This is a really simple way to take care of your cuticles.

FaceTime (As Nature Intended)
A Facemask

You will need:

- 1 tablespoon rice flour,
- A pinch of turmeric,
- 1 teaspoon honey,
- 1 tablespoon of mashed banana, tomato, orange, papaya or pineapple.

All you have to do is:

- Blend the dry ingredients together,
- Drizzle some honey to form a paste-like consistency,
- Finish off by adding your preferred fruit in.

Use it like any other face mask. It is a great natural alternative that will provide you with supple, hydrated skin. However, make sure you only use this mask once a week.

Scrubby, Dubby, Doo
A Body Scrub

You will need:

- ½ cup coconut milk or regular milk,
- ½ cup granulated sugar,
- 1–2 teaspoon coconut oil (as an additional moisturizing agent).

This one is really simple. All you need to do is:

- Take the sugar in a small bowl,
- Pour the milk and oil in and mix well,
- Stir until all the ingredients are mixed together,
- Apply to your body, scrubbing in circular motions.

This is great for exfoliating your whole body!

___

Ban-Druff
Anti-dandruff Therapy

You will need:

- ½ glass apple cider vinegar,
- ½ glass warm water.

Follow this method twice a week to get rid of dandruff:

- Mix the ingredients in a spray bottle,
- Spray the mixture on to your hair (ensure that it gets on to your scalp),
- Let it sit for 15 minutes,
- Wash it off and style as usual.

___

< 28 >

Squeaky Clean
Sanitizer

You will need:

- Rubbing alcohol,
- Aloe vera gel.

You can take care of your personal hygiene even after you have left your bathroom:

- Mix ¾ quarters of a cup of rubbing alcohol with ½ cup of aloe vera gel.

This is a really simple way to keep the germs away!

Make Your Own Cloth Pad[53]

A lot of small businesses across the country have started manufacturing eco-friendly cloth pads that are reusable, but in case you want to make one at home, follow the instructions below (this can also be used in combination with a menstrual cup, as a panty liner):

You will need:

- Flannel (roughly 50 cm in two colours will make you a couple of pads),
- Sewing machine,
- Thread,
- Measuring tape,
- Scissors,
- Cloth pad liner pattern or another cloth pad as a stencil or reference,

- An old towel,
- Snaps or tic-toc buttons.

Follow these instructions to make your own cloth pad:

- To make the outer body of the pad, trace the pad liner pattern, or your reference piece, and cut out two pieces. You can use a pen or pencil to trace it out,
- Make sure you add more length to the wings in the flannel if you are using a single-use commercial pad for reference. This is because the wings on these overlap when fastening the snaps or tic-toc buttons,
- Next is what goes inside the pad. Take an old towel that is still usable and cut out some rectangular shapes. Ensure that you give them curved edges,
- These need to be 2.5 cm shorter than the flannel cut-outs. Place the towel liner (centred) between these. You can place more than one based on your preference,
- You can keep the layers in place using one or more pins in case you're worried about things moving around,
- Now comes the stitching part. You can either sew the pads by hand or use a sewing machine. Either way, stitch two channels in the centre to fasten the towel on to both the flannel pieces,
- Next, carefully stitch all around your pad, sticking as close to the edge as you can, and then backstitch at the beginning and end. Make sure you align and pinch the layers so that they stay put as you work around them,
- Now you need to fix the snaps or tic-toc buttons on the wings to hold the pad in place,
- Make sure you place the male snap/button on the upper side of one wing and the female snap on the lower side of the other wing.

Your reusable pad is now ready! If a DIY version isn't for you, keep in mind that there are many businesses providing eco-friendly pads that you could use. Alternatively, you could try a menstrual cup to see if it is comfortable for you.

reuse old clothes
to make bags

pick ethical
or slow fashion

buy minimal clothing
and high quality
so it lasts

tailor your
own clothes

ZERO-WASTE
CLOSET

avoid compulsive
buying

leave the shoe box
at the store while
buying shoes

clothes swap
with family/friends
or at stores

STORES

donate
clothes

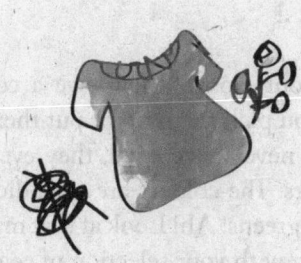

CHAPTER 2

CLOSET

'The staggering statistics of the quantity of clothes that end up in landfills is not news—in fact, what is new information is that landfills are brimming with so much urban waste that, by 2050, India is reportedly going to need a landfill that's the size of its capital, New Delhi.'[1]

The Perfect Life . . .?

Your clothes fill your closet. They have a certain smell, feel and sound when you pick them up or put them away. Perhaps, although you may never have tried, they even have a certain taste. They are yours. The colours are splendid! There are dark reds and absorbing greens! Ah! Look at that mauve top hanging in the middle along with your selection of coats.

You know that your coats will come in handy when it gets colder. You have already researched the colours that you want to pair with them, all to be bought during the upcoming sales—a little brown, a dash of charcoal and, a surprise to yourself, the suggestion is to go for dots instead of stripes! You didn't see that coming at all. It was certainly fortunate that you didn't buy anything at the last sale. The store had mountain loads of apparel, shoes, belts, hats, scarves and a rack full of handkerchiefs, too. How silly! Making clothes that don't fit the season, you thought for most of the items you viewed. There were those couple of things that you almost bought. Seemed to be your lucky day, didn't it? Also, you had seen what you wanted to buy for the latest season, but there was nothing to be found in that store.

At that moment, you remember a conversation with your grandparents from the previous day. How uninformed they were! You most definitely were not talking about spring, summer, winter and autumn. No, not at all. There are so many seasons for fashion in the year, more than four for sure. Silly old people who don't understand fashion, you judge.

Your eyes cast your mind back to the present. You pick up the shoes that you need on that day. You'll wear these for the event today and the function next month, but not after that, which seems perfectly suitable. In comparison, the clothes that you have chosen, you notice as you stare at yourself in the mirror, seem a little tired because you had to wear them for the reunion—both your undergraduate and postgraduate reunions—at the beginning of the month, as well as when you went for dinner with your family the week before. 'Oh well!' you say as you shrug and smile at how you look in the mirror. These things will be out of vogue next week anyway, just in time to collect the next season's necessities. With a cheeky smile, you close your cupboard and walk out the door, wondering for a fleeting moment if you should change into something that you haven't worn quite as often.

What Is In Your Waste?

'In a linear fashion model, it's estimated that 73 per cent of all our clothes end up in landfills for various reasons, like the lack of collection systems and ineffective redistribution.'[2]

'Reflecting on it now, it's funny that I did not have a profound sense of what ethical and sustainable fashion was until I returned to India in 2015. Even as a child, I remember that my mum ran an organic baby clothing workshop in our garage. It was my playground in many ways. The women who worked there used to make clothes for my dolls and show me how they used their sewing machines. They let me get a sense of the material that they were using and dressed me in clothes that matched my sisters', too. Yet, I was never truly aware of what all of that meant. I didn't realize how fortunate I was to have an opportunity to view the supply chain of a sustainable fashion business and gain a complete understanding of the effort that went into it.

In hindsight, I could not believe that I was simply viewing all this without really comprehending the reasons behind it. Even when I was away for years at college, I had always chosen to buy high-end fashion apparel at rock-bottom prices. It was only when I returned to India that I realized I had become completely detached from the entire process of production of clothes. Through the choices I made and the experiences I had, I took baby steps and learnt how to become increasingly aware of my fashion choices. I suspect you will witness the same results for yourself once you begin examining your own habits.'

NOW IT'S YOUR TURN TO ASSESS!

Try out Activity #1 with your current knowledge. View it as an initial way to see how much you know. Once you learn more as we go forward in this chapter, you can return to it in order to see how much your knowledge has developed.

< 36 >

ASSESS YOUR
CLOSET WASTE:
ACTIVITY #1

Follow the example sheet below by choosing a few of your own products to assess.

	Example A	Example B	Example C
Is the material natural or synthetic?	✓	✗	✓
Is the supply chain of the producer transparent?	✗	✓	✗
Does the product contain chemicals/ingredients that can harm the environment?	✗	✗	✓
Is the product designed to last or to be thrown out?	✓	✓	✗

- If yes: share the solutions with friends and family.
- If no: research options.

Use the ideas from this sample to assess your waste. You can draw your own sheet based on this and create other questions for this assessment based on your needs.

What Resources Are Available?

'There were two "aha moments" that really made me transition to a position where I understood why ethical and sustainable fashion is important. The first was with SELCO Foundation, when I was working in north Karnataka. During this time, I had the opportunity to visit villages and watch the local women work on handlooms and sewing machines. It reminded me so much of my childhood! In those days, weeks and months that I got to know them, I felt more and more of an attachment with what they did and why they did it.

Just like so many things in India, the textile industry is an art. Every weave, every stitch is done with a purpose. Every person I met was creating something to tell a story, something that they really cared about. That was my first major wake-up call.

The second was while watching the documentary *True Cost*. It highlighted the effects of fast fashion on people and the environment. The situation blew my mind. It helped to put it all into perspective. It humanized the creators of my clothing and the welfare of their communities, it made me question the synthetic dyes used in my clothing and, as a result, the waterways being polluted. In short, the documentary made me re-evaluate my fashion choices.

The move from viewing a situation blindly to really understanding it from the human and environmental costs was profound. It changed everything about how I perceived the clothes I was wearing. I felt so much more attached to the saris that I had purchased during my fieldwork in the rural areas. They are items that I will cherish for the rest of my life and pass down to the next generation.

Many of our traditions provide us with wonderful resources, like saris, which are essentially one long piece of material that can be fashioned into over a hundred different styles![3] They're items that people buy and keep for the rest of their lives partly because of its multiple functions, like the ability to be draped in different styles while constantly evolving and telling a story. This flexibility, I've come to realize, is something that clothes bought in fast fashion consumerism do not allow.

Fast fashion is constantly trying to size you up and categorize you into small, medium, large or extra-large sizes, all the while being made out of material that will not last you even months. And that's a concept that saris do not restrict you to.

It's wonderful and inspiring to witness the traditional side of India and all it can teach us about sustainability, from traditional garments people hold onto for a lifetime to clothes swapping among family members. India is "India" in so many wonderful ways. It has a lot to teach us. I think, sometimes, all we have to do is wake up a little from simply viewing a situation to being aware of the ins and outs of each woven thread.'

TRY THIS ACTIVITY OUT!

In this chapter there are a number of suggestions that you can use for Activity #2. You can either read ahead and return with new knowledge or assess yourself now before learning a little more and returning to this exercise later.

ASSESS YOUR
CLOSET RESOURCES:
ACTIVITY #2

Visit a clothes swap in person/online or turn old clothes into something new. Use the suggestions in the guide book or research to find more. Fill in each section of this activity to record your achievements.

The resources that I have learnt about are:

 'Exchange Room' Bangalore clothes swap.

 'This For That' online fashion and beauty swap.

 How to change old shirts into a brand new quilt.

Draw a picture from your activity to show friends and family:

The resources needed for this activity are:

 Details on the clothes swap.

 I downloaded the online clothes swap app.

 Old shirts, sewing equipment

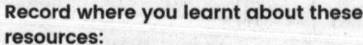

Record where you learnt about these resources:

 The guide book's 'Zero-Waste Clothes Swap' section and 'Zero-Waste Library'.

Record who you shared your success with:

 My friends! They came to the clothes swap with me, it was so much fun.

Use the above ideas to assess your resources and then create your own activity sheet.

Moving Towards a More Sustainable Lifestyle

'My first clothes swap was when I was very young, when I got my older sisters' hand-me-downs. It is something that everyone around me used to do. I had almost forgotten about it when I returned to India, but during the first year when I was learning about the fashion world, I visited a small organization that was running a clothes swap for anyone who wanted to attend. Honestly, though, it wasn't good. It was really small and there was not a great selection. I remember browsing through the clothes and not finding anything that was my style. However, I'm really glad that things have changed dramatically since those days.

I've been attending clothes swapping events for over three years now. The one that I really like is nothing like the first one I visited. Nowadays, everyone is donating. There is an insane amount of choice available! At an event I attended in Bengaluru late in 2019, there were over 100 people in attendance. It has become a community run by volunteers where you meet like-minded people. The growing popularity of clothes swaps represents a shift in culture and highlights the changing value-based consumption patterns of this generation. Millennials are increasingly aligning their choices with consumption patterns. A massive fast fashion brand like Forever 21 filing for bankruptcy is a testimony to this fact.[4] The whole concept is helping people become more aware of the environmental and human impacts of the fashion world. It's almost like you have a new wardrobe to choose from every time! Given this, clothes swaps are growing all over the country, marking a big step up from the days when we used to share clothes within a four-people family only. Every swap is a new layer to the story of that piece of clothing. It is important to remember and cherish that while attending these events. It may be a little daunting to begin with, but these days they are so well established and welcoming that anyone who attends will get the hang of it straight away.

< 41 >

There are a number of organizations I want to recommend to you who really inspire me in this fashion realm. Their names are listed below for you to learn more about them.'

ZERO—WASTE CLOTHES SWAPS

Here are seven different clothes swaps in some of India's larger cities. There are less-well known businesses in some of the smaller cities, too. All you have to do is inquire online (or ask some of your tech-savvy friends and family):

- **Switcheroo** is an event hosted in Hyderabad by the Global Shapers Community, a global 'network of young people driving dialogue, action and change'. They swap menswear, womenswear and accessories, too. They accept anything in great condition, with a view to promote sustainable use of products.[5, 6]

- **Eco Bazaar** is an initiative by Nidarshana Saikia Das and Aruna Nayagam. They run a monthly event with a clothes swap as part of each edition. They have run swaps in Hyderabad for both women and men, making it a fantastic way to promote sustainable fashion across genders.[7]

- **Exchange Room** is a terrific eco-friendly clothes swap. The community that started with ten participants in July 2014 now has thousands of Instagram followers. This event is regularly hosted in Bengaluru.[8]

- **GreenStitched** began in 2015 with a goal to inspire sustainable fashion through talks, films and other promotional tools. The swaps are run under SwapStiched with multiple events having already been organized in Bengaluru, and one in Mumbai, in collaboration with Fair Trunk.[9]

- **Apparent Club** is a peer-to-peer app that focuses on lending and renting clothes. They are based in Mumbai and building an online presence. They provide consumers with clothes and information on the inherent dangers of fast fashion. Their aim is to increase the life of products by creating a circular economy for the apparel they rent.[10, 11]
- **This For That** began as a series of clothes swap parties in Delhi and the National Capital Region but has evolved to become an app-based organization. This means that it is now available across India to help women swap fashion and beauty products. As of now, they have 4000 to 5000 people using the app each month.[12]
- **Everwards** is a clothes swap event that is making great inroads into Chennai. The organization sells the clothes rather than swapping them but for an incredible 10 per cent of MRP. They take special care of the products to ensure that the pre-loved clothes and jewellery are in pristine condition for the next person. The organization is aiming to encourage a shift in mindset, away from impulse buying and towards conscious consumerism.[13]

'There are so many environmental champions both in India and across the world these days. Honestly, it's an exciting time to be alive and involved in this movement. You can engage in many different ways, some of which are as easy to do as the tips listed below and others where you will learn a little more about the supply chain of all products, including clothes. The illustrations on the next couple of pages highlight the beginning of the "life span" of all products. There are many materials and people involved; we'll keep developing these concepts in the coming chapters as well. This way we will gain an important visual lesson of the difference between a wasteful method and an earth-friendly one.'

ZERO-WASTE TIPS AND TRICKS

- Pick ethical or slow fashion,
- Clothes swap with family or friends, or in clothes swap stores,
- Reuse your clothes, e.g., cut your old clothes into small pieces to use as rags around the house,
- Leave the shoe box at the store when buying shoes,
- Donate clothes,
- Buy minimal items of high quality so that they last,
- Avoid compulsive buying,
- Tailor your own clothes.

< 44 >

CLOSET - LINEAR EXTRACTION

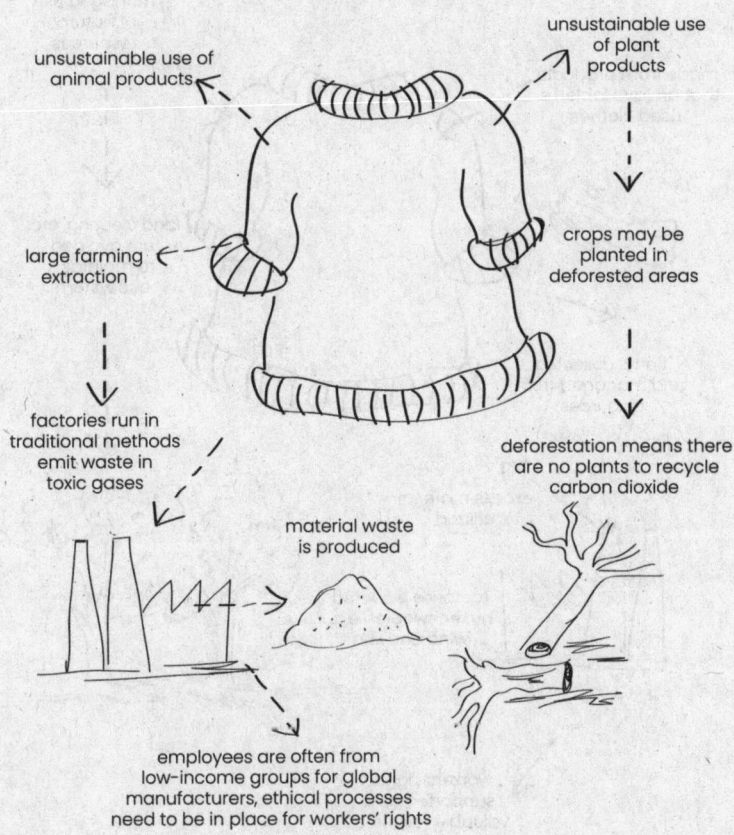

unsustainable use of
animal products

unsustainable use
of plant
products

large farming
extraction

crops may be
planted in
deforested areas

factories run in
traditional methods
emit waste in
toxic gases

deforestation means there
are no plants to recycle
carbon dioxide

material waste
is produced

employees are often from
low-income groups for global
manufacturers, ethical processes
need to be in place for workers' rights

CLOSET - CIRCULAR CREATION

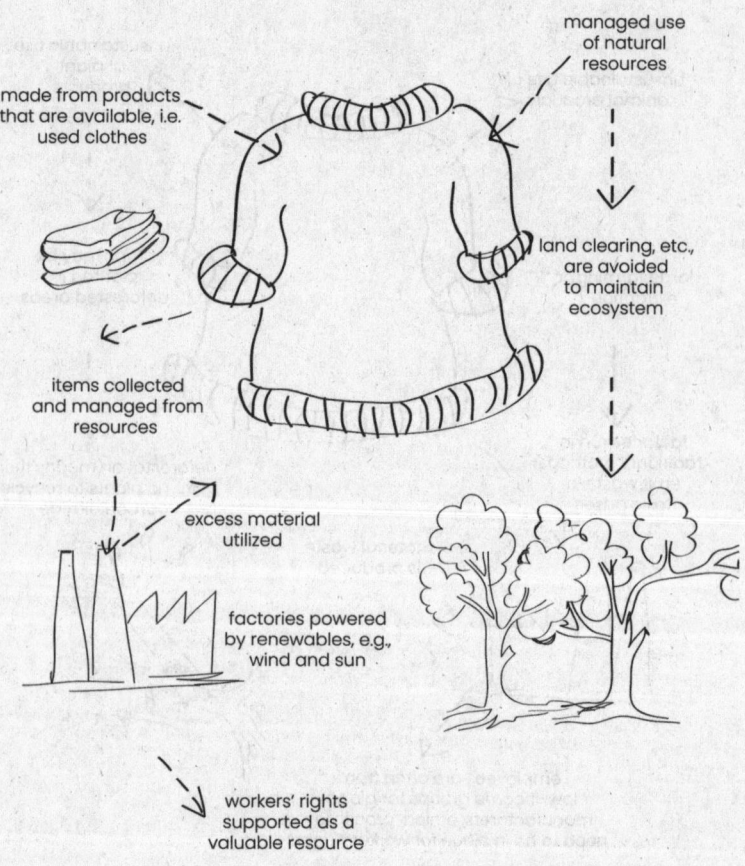

managed use of natural resources

made from products that are available, i.e. used clothes

land clearing, etc., are avoided to maintain ecosystem

items collected and managed from resources

excess material utilized

factories powered by renewables, e.g., wind and sun

workers' rights supported as a valuable resource

Why Is It Harmful?

> 'The choices that we make with our clothes have far-reaching impacts. I'd really encourage you to learn how our choices are connected to things close to home and faraway, from the time a shirt is made, for example, until the time you no longer want to wear it.
>
> You now have a chance to read about how the supply chain of consumer products and material waste is interconnected with the buy-and-sell world we live in. I'll touch base with you in the final section to see what you have learnt of these lessons.'

There may have been times in your life when you have bought an item of clothing, returned home and not liked the look, size, shape, colour or another attribute for a particular reason. On a singular level, this may not look like a big issue if your shirts or trousers, shoes or scarves were slightly at odds with how you originally viewed them. Yet, if this was to be looked at globally, considering the scale of fashion marketing and what that does to the mindsets of consumers worldwide, it has an all-encompassing impact.

The Supply Chain

'[As much as] 20,000 litres is the amount of water needed to produce one kilogram of cotton; equivalent to a single T-shirt and a pair of jeans.'[14]

Fast fashion is the production of clothes that look like high-quality products at cheap rates,[15] where many standards are

< 47 >

often overlooked throughout the manufacturing process for profit. For example, working conditions of the employees are often disregarded, sometimes with dire consequences. On 24 April 2013, a garment factory based at Rana Plaza, an eight-storeyed building located on the outskirts of Dhaka, Bangladesh, collapsed. It killed 1134 people and left many more injured.[16] There had been complaints by employees about the building having cracks. Unfortunately, the larger fashion brands who operated the facility chose to focus on profits and mass production instead of addressing the structural issues with the building, highlighting a lack of responsibility and accountability[17] to the employees.[18]

The situation does not simply remain isolated to one location or one brand, nor does it look at the entire ecosystem.[19] If you were to assess all costs, for the overall impact of production and consumption, the added expense on the planet's resources would increase the price of a new item of clothing significantly. Viewing the practices of some large fashion retailers worldwide, which get daily shipments of new styles with certain retailers attempting to introduce up to 400 styles a week on their websites,[20] this additional expense is not calculated.

There are far too many items of clothing being produced to be utilized and therefore there is nothing else that can occur other than clothes ending up in landfills. At the same time, there are now far too many impacts on the planet for consumers and suppliers to stay naïve to the overall impacts of an unsustainable supply chain.

External Shocks

'While we have been encouraging an end to overconsumption for many years, we also know that in the face of this unexpected halt in manufacturing, it is the most vulnerable, lowest-paid people in the fashion supply chain that feel the worst effects.'[21]

Despite realizing the vulnerable positions that employees endure in locations such as Bangladesh, and many others,[22] the linear mass production model remains unchanged. Transitioning from a wasteful method to a resource-valuing system will often not occur overnight. Though there are positive signs, which will be noted below, the old methods remain, leaving individuals at or below the poverty line vulnerable to unexpected changes.

The COVID-19 pandemic[23] has highlighted this situation.[24] The global fashion industry 'typically pay their suppliers weeks, or even months, after delivery, rather than upon order'[25] which has led to a reported '1089 garment factories in Bangladesh [that] have had orders cancelled worth roughly $1.5 billion (USD)[26] due to the coronavirus outbreak . . . Many factories in Bangladesh have been shut down indefinitely. Some workers were given less than a month's salary as severance and many others received nothing at all.'[27] Similarly, across India, informally employed, minimum-wage workers are forecast to be without employment for upwards of half a year, without social security, savings or a readily available safety net.[28]

The human impact of an industry closing the doors is profound.[29] COVID-19 has had catastrophic impacts all over the globe, on a wide range of sectors,[30, 31] not just on fashion. The pandemic has presented every person on the planet with an extreme example of what could occur when business as usual cannot be achieved. The hope is that humanity learns from the crisis and implements better methods that protect people from unforeseen events. Textile workers[32] and others need this support. They need a stronger, more robust system. Fortunately, even before the pandemic, there were individuals and organizations, a couple of them listed below, who were coming together to implement new ways of production and consumption.[33]

< 49 >

Material Waste

'[As much as] 85 per cent of textiles go into landfills each year. That's enough to fill the Sydney Harbour annually.'[34]

The fashion industry produces 8 per cent of global carbon emissions, which is in excess of the combined total of international flights and maritime shipping.[35] This is a truly global issue, one that is contributed to by consumers all over the world. Some fashionistas change clothing styles up to fifty-two times a year![36] Therefore, it is not solely the manufacturers at fault for this waste, consumers, too, have been playing a key part.

The impact that fast fashion trends and marketing have on consumers is likely due to the invisibility of the wider costs at the cash register, with the true costs being borne by impoverished communities and the environment. The result is that over the past decade and a half, the fashion industry has doubled in production and, at the same time, clothes are being thrown away 40 per cent sooner.[37] Moreover, 73 per cent of the garments that end up in landfills are burned, adding to air pollution, while only 12 per cent are recycled into other materials such as stuffing for mattresses.[38]

On a positive note, some large companies such as Patagonia.[39] have been able to show the value of reusing material in a circular economy. This is emphasized by several independent experts such as the Ellen MacArthur Foundation[40] that is calling for new methods and a system level change with unprecedented levels of commitment, collaboration and innovation in the entire textile industry.[41] Further, a growing level of awareness about these issues is being recognized by consumers across the globe, namely those who are involved in Fashion Revolution,[42] a global consortium of people focused on moving fashion to more sustainable practices.

Yet, with waste from clothing amounting to a garbage truck worth of material (2625 kg) being burnt or poured into a landfill every second,[43] people all across the globe need to get involved. There is an urgent need to move to new methods where the materials are either reused and/or recycled.

What Can You See in Your Waste Now?

'It's pretty shocking to hear some of the environmental impacts in this chapter, isn't it? For these reasons, among many others discussed throughout the guidebook, it is vital to keep looking deeper. To the next step in the process, and the one after that, and so on, to really learn how you can make a difference.

We can all help to change the world, keep that in mind! A simple choice can have countless positive effects. Ultimately, it all comes down to you.

Enjoy the last section here. I'll see you in the next chapter, but remember to keep bringing the lessons that you're learning along with you.'

LOOK A LITTLE DEEPER WITH THIS ACTIVITY!

Activity #3 is a four-part process, on separate question sheets, that builds on everything you have learnt throughout this chapter. The importance of this exercise is to understand and become aware of the processes that can help you transition to a sustainable lifestyle.

©Bare Necessities Zero Waste Solutions Pvt Ltd

ASSESS YOUR
CLOSET WASTE:

What type of environmental impact do your products/services have?
Join the product/service to the problem using an arrow:

Product/Service:

Shirts

Shoes

Bags

Problem:

Unable to recycle
or decompose

Synthetic dye
polluting water

Plastic pollution

Use the ideas from this sample to assess your waste. The product/service
may relate to more than one problem. Fill in the lines below with your
products/services and environmental problems.

———————   ———————

———————   ———————

———————   ———————

———————   ———————

When thinking about the environmental impact, it is important to think of the number of areas that it could
affect. Start on a small scale and work your way out. First, think about what it means to the micro-
environment around you, then think larger and larger until you look at it from a global perspective. It will be
beneficial to undertake some research online or in a library.

ASSESS YOUR
CLOSET WASTE:

What system issues prevent change?
Join the product/service to the issue using an arrow:

Product/Service:

Shirts

Shoes

Bags

Issue:

No regulations

No affordable alternatives

Big businesses not accountable

Use the ideas from this sample to assess your waste. The product/service may relate to more than one issue. Fill in the lines below with your products/services and environmental issues.

_____   _____

_____   _____

_____   _____

_____   _____

_____   _____

A system is anything associated with the product or service that is interconnected with it, for example, the manufacturing unit that produces the product or the government office that provides the service. To learn more, conduct research online or in a library.

ASSESS YOUR
CLOSET WASTE:

What sustainable options are there to replace it?
Join the product/service to the solution using an arrow:

Product/Service:

Shirts

Shoes

Bags

Solution:

Make your own

Choose natural
fabrics

Shop from ethical
enterprises

Use the ideas from this sample to assess your waste. The
product/service may relate to more than one solution. Fill in the lines
below with your products/services and solutions.

_____   _____

_____   _____

_____   _____

_____   _____

_____   _____

A sustainable option is a product or service that will last longer and/or produce less waste. Think about
options such as products made from earth-friendly materials, or those that reduce waste through a supply
chain. To learn more, research online or in a library.

ASSESS YOUR
CLOSET WASTE:

How will you start using the sustainable option?
Join the product/service to the action using an arrow:

Product/Service:

Shirts

Shoes

Bags

Action:

Upcycle old shirts
into a new bag

Purchase from
ethical enterprises

Invest in long-
lasting footwear

Use the ideas from this sample to assess your waste. The
product/service may relate to more than one action. Fill in the lines below
with your products/services and actions.

_____   _____

_____   _____

_____   _____

_____   _____

_____   _____

This last step is all up to you. Make your choice, know the benefits and live a zero-waste lifestyle. You are more
likely to succeed with support from your friends and family.

ZERO—WASTE LIBRARY

A Quilt of T-shirts[44]

If you're a collector of T-shirts and find yourself with drawers full of memories that you don't want to let go of, but also really don't want to wear any more, consider giving them some practical value by upcycling them into a T-shirt quilt. I did this with all my college T-shirts that I was too sentimental to get rid of.

T-shirt quilts are a fantastic way to preserve memories or create a truly one-of-a-kind gift. With just a little bit of prep work (and a whole lot of T-shirts), you can give some of your treasured clothes a second life.

Where do you begin?

- First, pick out the T-shirts you have decided to use,
- Second, divide them into themes.

Here's a simple guide to help you estimate how many T-shirts you might need:

Quilt Size	Shirts Needed
Lap:	16
Twin:	24
Full:	30
Queen:	49
King:	64

Step 1

Determine the fabric needed for your T-shirt quilt. To make a full-size quilt (150 cm x 182 cm) you will need the following:

- 30 T-shirts. It works best with graphics that are 22 cm or less wide and 25 cm or less in length. If you use the front and back of a single T-shirt, you will need fewer,
- 1325 cm Pellon Fusible Featherweight (911 FF),
- 152 cm x 182 cm batting,
- 68 cm binding,
- 152 cm x 182 cm backing.

Pro tip: It's best to wash and dry your T-shirts before beginning. Shirts that are a little worn out are fine, but avoid using those with holes or rips.

Step 2

Cut your T-shirts:

- Using your rotary cutter, cut thirty 40 cm x 40 cm squares from the light-weight fusible interfacing,
- Next, cut the shoulder seam, sleeve seams and side seams of each T-shirt,
- Then, position the fusible on the back of each T-shirt, paying attention to the graphics on the front of the T-shirt,
- Press the fusible interfacing on to the back of each piece according to the manufacturer's directions.

Pro tip: Do not place your iron directly on the T-shirt graphics because the ink may melt. Press the back of each T-shirt.

Now it's time to cut out your squares:

Pro tip: To help you, AccuQuilt has two Studio dies that make cutting the T-shirts a breeze. The Studio T-shirt Quilt Squares (12 inches and 16 inches) feature registration pins for perfect alignment.

- If you are working with adult-sized T-shirts, your block can be up to 40 cm,
- For childrens' or junior-sized T-shirts, the 30 cm square is best,
- Cutting one T-shirt at a time is recommended. Pay careful attention to the placement of the design,
- Feel free to cut more keeping in mind that fusible web counts as half a layer of fabric and that T-shirts are thicker than typical quilting fabric,
- Place your T-shirt on the die (stencil), positioning the graphics in the centre and cut around the edges of the die,
- Cut thirty squares,
- You can also do this by hand, use a square pattern as a reference and trace them out on each T-shirt with a pencil, and then cut out the squares.

Step 3

Sew your T-shirts together:

- Lay out your squares to find a pleasing configuration,
- Remember that the top, bottom and sides will be trimmed later,
- Now it's time to sew your rows,
- Take the first and second squares from the first row of your quilt (on the graphic sides). Place them together and sew the left side,
- Next, put the third square on the second square and sew the right edge together,
- Continue until you've sewn an entire row,
- Press the open seams to reduce bulk,
- Once all your rows are ready, it's time to put them together,
- Pin the graphic sides of row 1 and row 2 together, aligning the seams. Then sew the top seam,
- Press the seam open,
- Repeat until all your rows are sewn together.

Step 4

Finishing your T-shirt quilt:

- Use a 15 cm x 60 cm ruler and a rotary cutter to square-up the quilt along the top, bottom and sides,
- Layer batting between the quilt top and backing,
- Next, pin or baste the layers together,
- Finally, machine-sew with your desired design. Add binding and join the corners.

You're done! Now you can enjoy your favourite T-shirts in a whole new way!

A Face Mask Made From Your Old Clothes[45]

You will need:

- Tightly knit cotton material (this can be a T-shirt, scarf or something similar). The fabric should be large enough to fold several times in order to cover your nose and mouth,
- Two rubber bands or hair ties.

To assemble this mask that can protect you from dust, or prevent the spray from a cough or sneeze from entering your respiratory system, follow the instructions (based on a T-shirt as reference) and use the same dimensions if you are working with a scarf or any other item):

- Cut around 17–20 cm of the fabric from the bottom of the T-shirt. Cut horizontally to get a long strip,
- Lay the material out flat in front of you and turn it 90 degrees so that what used to be the bottom hem of the T-shirt (it's usually double stitched) is facing left or right,

- Fold it from the bottom to the centre, and from the top to the centre,
- Repeat this step a second time,
- Loop a rubber band or hair tie around each end (left and right), leaving a few inches of fabric, so that each side looks like a candy/chocolate wrapper,
- Fold the excess material over the band, with each side meeting in the middle, adding another layer to the mask,
- Put a band over each ear, making sure the material fits snugly on your face.

The pressure on your face should hold the material and rubber bands in place.

A DIY Bandana-and-Coffee-Filter Mask[46]

This mask uses a coffee filter, placed in the middle of a layered bandana, to provide protection.

You will need:

- A bandana,
- Conical coffee filter,
- Two rubber bands or hair ties.

To assemble this mask, follow the instructions:

- Fold the bandana square in half,
- Cut the coffee filter horizontally across the middle,
- Place the wide section of the filter in the middle of the folded bandana,
- Fold the bottom of the bandana up to the middle, covering the filter, and fold the top down over it again,

- Loop a rubber band or hair tie around each end (left and right), leaving a few inches of fabric, so that each side looks like a candy/chocolate wrapper,
- Fold the sides over the band so that they meet in the middle. Tuck them together.
- Put a band over each ear, making sure the material fits snugly on your face.

The pressure on your face should hold the material and rubber bands in place.

Sew Your Own Face Mask[47]

This mask is more difficult to make, but it may feel more comfortable and will last longer.

You will need:

- Two 25 cm x 15 cm rectangles of tightly woven cotton fabric. You can use sheets or quilting fabric, or even a T-shirt,
- Two 154 cm pieces of elastic/rubber bands/hair ties/string/cloth strips,
- A sewing machine,
- Needle and thread.

To assemble this super-comfortable face mask, follow the instructions:

- Stack the two rectangles of fabric together,
- Fold the 25 cm sides 0.5 cm down and sew them together,
- Fold the 15 cm sides 0.5 cm over and sew at the edge, leaving a small gap for the elastic to loop in,
- Thread the elastic through the gap,

- Tie or sew the ends together,
- Tuck the knots inside the opening. Gather the short sides together and stitch the elastic into place.

You're all done! This is a fantastic face mask that may be more comfortable than the other two.

The Soles on Your Feet

There are shoes all over the planet: some are discarded, some are still being worn and many are sitting in landfills. To help address this wasteful situation, there are organizations around the world focusing on upcycling footwear. One such organization is Greensole, which is based in Mumbai. It aims 'to contribute to social good by creating a self-sustaining infrastructure that facilitates the provision of the basic necessity of footwear to everyone, forever, environmental good, by refurbishing discarded shoes with zero carbon footprint and economic good by giving employment to refurbish shoes'.

There are others, too, all around the world. With some research online, you can learn more about how society can repurpose footwear.

Recycled Towel Bath Mat[48]

To reuse your old towels that are no longer as good as they once were, try to repurpose the material.

You will need:

- 3 to 4 bath towels (choosing different colours will make the design more spectacular),
- Cutting mat,
- Rotary cutter,

< 62 >

- Fabric scissors,
- Pins,
- A needle,
- Thread.

Follow the twists and turns of the following instructions:

- Cut the towels into 7 cm-wide strips,
- Line up three strips (cut off the border of the towels if necessary),
- Pin and sew them together,
- Remove the pin,
- Fold the sides of one strip in towards the centre and then fold it in half,
- Pin to secure and repeat the entire length of the strip,
- Repeat with the other two strips,
- Twist the three strips together to create a colourful effect. Remove the pins as you go along,
- Once you have reached the end, sew on more strips and continue until you run out,
- Sew the end to secure,
- Twist the long rope into a coil and securely sew together.

There you go! You've made a really unique bath mat. Well done!

Call Them Nappies, Call Them Diapers . . . But Let's Make Them Sustainable!

Here are a couple of great Indian-based businesses that provide options to buy nappies/diapers for your little one. The best part is that they won't harm the planet like single-use plastic products do.

- **Bumpadum Diapers**: Bumpadum provides next-generation cloth diapers and accessories that are designed and made entirely in India.[49]

- **Superbottoms**: The team of all-moms at Superbottoms is passionate about earth-friendly ideas and gentle pro-baby parenting. All their products are made with love and have personal values and beliefs wrapped up in every stitch.[50]

Tote Bag

You will need:

- 1 old shirt,
- Scissors.

This is an amazing technique for reusing clothes, avoiding single-use bags at the shop and having some fun as well. It's a little tricky at first so follow the method and keep trying until you succeed. Take your time; you'll have a great story about your bag to tell everyone! The process is:

- Cut the shirt sleeves,
- Cut the neck out to make a deep 'U',
- Cut slits or a fringe at the base,
- Cut slits from the bottom of the shirt up to the line marking the bottom of your bag,
- You'll want to cut both the front and back layers together because they need to match for the next step,
- I cut the slits about 3/4 to 1 inch apart,
- Tie fringe or knots.

Okay, now this is going to sound really complicated, but it's not once you get the hang of it, I promise.

- Take the first pair of fringe and tie it into a knot. Then tie two more pairs,

< 64 >

- Now, if you lift your bag you'll see that although the pairs are pulling the bag together, there's a hole between each pair.

This next step will close those holes:

- Grab one strand from the middle set (the one with the arrow pointing left) and tie it in a knot with one of the strands on the left set,
- Take the other strand from the middle set (the one with the arrow pointing right) and tie it in a knot with one of the strands on the right set,
- Take the remaining strand on the right set and tie it to the next set of strands,
- Continue on each strand, back and forth, until all the strands are tied,
- Now turn your T-shirt right side out again and voila, you're done!

This is a really amazing technique. I hope you have fun with it!

Run Your Own Clothes Swap[51]

- Choose a space to host the event, such as your apartment, your office, your neighbourhood, your city,
- Pick an audience,
- Find a venue. This can be a virtual marketplace, too,
- Choose a method, for instance, will it be point-based, barter or monetary,
- Invite everyone. You can use social media to help spread the word,
- Get volunteers to help organize the event,
- Host your swap,
- Donate if you have spare items.

< 65 >

compost your
food waste

ZERO-WASTE
KITCHEN

bring reusable bags
to the shops and
buy fruits and
vegetables that are
package-free

reuse and
reinvent
your leftovers

buy milk
in reusable
containers

use steel utensils
and avoid non-stick
cookware

use reusable containers
for your ingredients

buy local
and buy in
bulk

use reusable
cloths and organic
scrubbers

CHAPTER 3

KITCHEN

'[As much as] 40 per cent of the food produced in India is either lost or wasted. This food wastage, however, isn't limited to one level alone but perforates through every stage; from harvesting, processing, packaging and transporting to the end stage of consumption.'[1]

The Perfect Life . . .?

Your grocery bags are plopped down with a soft rustle and thump on to your kitchen countertop. You extract your fruit and vegetables first, which were carefully placed on top of the other items so that they are not bruised. Each item is held separately within its own bag. Well, you did not want to have them mixed up. Even your potatoes, carrots and turnips are in separate bags. You place them on the left of your counter and throw the individual plastic bags away. Next, your fruits are placed into a bowl and each bag that held them within the larger grocery bags is deposited into your garbage bin.

You place many of the items you just bought into a cupboard that is already full. You notice that there are a few items that have crossed their expiry date and toss them into the bin, too. You sniff the milk that was in the fridge and make the same judgement call. The liquid is poured down the drain before the container is put away into the bin, too.

You have planned a large dinner with friends and family tonight. It's time to prepare. You have the rice ready. You knead the dough for your naan. You cut your vegetables. You wash your fruits as well, ensuring that all the chemicals that helped them grow at this time of year are washed off. Then you proceed to arrange them purposefully on a plate. You pull the condiments out of the fridge, emptying each and every jar. From the chutneys to the nuts to dahi (or yoghurt), you prepare everything as though you are a professional chef, maybe even better than one! You think to yourself that there is no way that any chef would put this much love and attention into the meal that they prepare.

Several hours later, you are done. The black bag in the bin is overflowing with vegetable peels, the cores of the fruit and many other tins and containers. You lift the bag and can identify the shapes of the things you discarded. There are the cardboard boxes that held the flour, the glasses that held the chutney, the firm plastic shape that once held the kilograms of rice and the softer plastic that had been home to the nuts and dried fruits. You walk out of the door and place the bag by the road, just far enough from the door to ensure that your guests do not notice. You hope that the cows, goats or dogs don't tear into the bag before the party.

Back inside, you clean the kitchen with the strongest citrus-smelling surface cleaners and detergents available. They are almost finished, so you pour the last bits down the drain that had earlier swallowed the sour milk. You smile as you place another bin liner into the dustbin and toss the bottles in. You hear the doorbell. You are ready just in time to feed your guests some delicious food.

What Is in Your Waste?

'The world's second most populous country needs to reduce its food wastage to feed the 194 million Indians who go hungry daily.'[2]

> 'When you think about it, we share an amazing attachment and intimacy with food in India. We grow food with our hands and we eat with our hands. It is an extra level of affinity. Despite that, and the fact that I love eating food, the kitchen was not an area that I knew much about when I started my zero-waste journey. I lacked a well-rounded understanding of the attachment and intimacy of eating and enjoying food as an Indian in India.

< 69 >

A big moment in my journey came when I was visiting a Bengaluru-based social business that specializes in composting, Daily Dump. They had been operating for ten years when I met them, and they gave me my first insights into composting and why it is important. The knowledge I gained from the founder of the business, Poonam Bir Kasturi, is among the top-five most important experiences in my transition to a sustainable lifestyle.

In India, approximately 60 per cent of waste is organic,[3] which means that if you consider a banana peel, it is no longer waste if managed in an environmentally conscious way. If it is placed into a composter, it will degrade naturally. At that point, it becomes nutrient-rich soil for plants and other life forms. Of course, the organic products need to be segregated from other material, but apart from that all you need is a composter and a little time.

I bought an aerobic composter (this composting method uses oxygen to help decompose the organic products), which is commonly known as a *khamba*. It is made from terracotta clay and handcrafted by local Indian artisans or potters. My banana peel and the other organic waste I placed inside it decomposed this way:

- Place a small amount of remix powder inside (refined and decomposed compost with a few natural microbes to help speed the decomposition. Microbes are not a necessity, but they do reduce the waiting time if you want to start working in your garden faster with nutrient-rich soil that you have helped produce),
- Place the banana peel and other organic products inside, on top of the layer of powder,
- Sprinkle a small amount of remix powder on top (this controls smells, too!) and place the lid of the composter back on,
- Repeat the next day with more organic products,
- In just 4–6 weeks of layering, the composter will be full. The material will have decomposed, giving you nutrient-rich soil for the garden.

I've got some other really important things to share with you too. But first I want you to look at your kitchen waste.'

NOW IT'S YOUR TURN TO ASSESS!

Try out Activity #1 with your current knowledge. View this activity to see how much you know. Once you have learnt more as we move forward in this chapter, you can return to it in order to see how much your knowledge has developed.

< 71 >

©Bare Necessities Zero Waste Solutions Pvt Ltd

ASSESS YOUR
KITCHEN WASTE:
ACTIVITY #1

Follow the example sheet below by choosing a few
of your own products to assess.

	Example A	**Example B**	**Example C**
Can this product be bought in a less harmful way, i.e. less packaging?			
Is the product grown organically?			
Does the product contain harmful chemicals, either sprayed on or as an ingredient?			
Is the container recyclable or reusable?			

- If yes: share the solutions with friends and family.
- If no: research options.

Use the ideas from this sample to assess your waste. You can draw your own sheet based on this and create other questions for this assessment based on your needs.

What Resources Are Available?

'Buying the composter was definitely a turning point for me, but changing my kitchen into a zero-waste area was not easy or immediate. Like many Indians, I live with my family members. I learnt quickly that just purchasing a composter did not mean that they were going to change their practices. This in itself was a big learning curve about what it means to try and live a life congruent with my own values. I saw that everyone around me had their own reasons for following their own practices and that forcing ideas upon them would not be an effective way to promote change. This was an enlightening moment, but also a huge stumbling block. For a while, I was confused about what to do.

At the same time, I was bringing home bags full of peanuts, millet and other produce from my fieldwork in northern Karnataka. This provided me with an interesting perspective of what life used to be like in India before single-use items invaded our economies and lifestyles.

Bazaars, for me, are a quintessential part of India, something I cherish from my childhood. But, I think, we have moved away from this concept. Shopping at bazaars means that there is not a pre-packaged amount of food that you are required to buy, instead you can select what you want and how much you want, without the unnecessary bags and packaging that we see in supermarkets. Going to a bazaar is not only inherent to our culture, but also shows all of us how we once lived.

Reflecting on the situation, I realized that before I left for countries that had high volumes of takeaway orders, all I had witnessed were local market purchases of large items, like the millet or peanuts I brought back home from the villages. Yet, in the time that I was away, the advent of the food services industry and the increase in the number of supermarket chains in India began. This appeared to be a mirror image of what I had viewed in the countries I had studied in, but on a much bigger scale.

This situation was really noticeable in Bengaluru due to a waste management system that became outdated when the city started experiencing a huge population growth post the IT boom. I understood that this changed system would also have a huge effect on the farmers I was visiting if the largest companies decided that they no longer wanted to sell millets, for example, or if someone else decided to sell their farm stock at a cheaper rate. Such selective profit and greed-driven changes brought about by large corporate retailers could lead to less diversity of farm produce, leaving the farmers even poorer (while also impacting the fragility of the ecosystem). But, in reality, what we fail to realize is that our farmers carry a wealth of knowledge—and profit-driven discrimination leads to them becoming poorer socially as well. Knowing that such discrimination occurs simply for profitability was extremely disheartening.

In many ways, I see it as a huge wake-up call for all developing countries. While there are some benefits of modelling developed nations, there is also a need to review whether these methods work locally and what impact their implementation can have on the people.

Meanwhile, Bengaluru found itself in a perplexing situation. It was a city that was expanding, with food delivery companies that now mimicked the West. The people I lived with were reluctant to change in the kitchen. Farmlands had monoculture crops. Villages did not benefit from a growth in consumerism. The situation felt out of control for some time. It wasn't until I realized that I needed to make more of a practical effort that I saw results. This meant that I needed to communicate in other ways with the people I shared a kitchen with about why our collective methods needed to change. The situation also made me realize that I had to focus on addressing the immediate things that I could change, establish them and then look at the larger issues that I cared about later.

Perhaps you will find similar things happening with you. The important thing is to try and work out the solution as you continue on your zero-waste journey.'

TRY THIS ACTIVITY OUT!

Throughout the chapter there are various suggestions that you can use for Activity #2. You can read ahead and return with new knowledge, or assess yourself now before learning a little more.

< 75 >

ASSESS YOUR
KITCHEN RESOURCES:
ACTIVITY #2

Organize a bin system for organic, recycling, toxic and reject items and/or start composting. You could also visit a composting and/or dry waste centre. Use the suggestions in the guide book or research to find more. Fill in each section of this activity to record your achievements.

The resources that I have learnt about are:

 I created a new bin area for all types of items.

 I took my recycling to a dry waste centre.

 I started composting.

Draw a picture from your activity to show friends and family:

The resources needed for this activity are:

 I reused a couple of containers that could hold all my items, like batteries, for proper disposal.

 I researched online about my nearest waste centres.

 I needed a composter, it is a really worthwhile purchase.

Record where you learnt about these resources:

 I learnt many things from the guide book, online and the composting centre.

Record who you shared your success with:

 My family! I live with them, we're all reducing waste together.

Use the above ideas to assess your resources and then create your own activity sheet.

Moving Towards a More Sustainable Lifestyle

'All the lessons that I was learning made me focus on the smaller areas based on which I could set a course of action. I began researching recipes incorporating local Indian millets, and I involved the people I lived with in the process, learning from them where I could. I found that they were more interested in the health and wellness aspects of shopping from organic sources, instead of buying from home-delivery companies, than the zero-waste side. It was a significant move in a positive direction for all of us.

We began buying bulk amounts of ingredients to make millets khichdi, granola mix for our breakfast and peanut butter, too, among other favourites. We also started placing bulk orders for fruits and vegetables from a small town called Ooty, situated in the hills of western Karnataka. This provided us with access to seasonal products of a wider variety compared to what could be found at a supermarket. This was really nostalgic for my mum. The quality of the food reminded her of the produce available when she grew up. For my yoga-practicing Ma, her motivation became focused on health and wellness rather than the zero-waste angle.

By focusing on the smaller areas that I could bring a significant change in, and by not taking a back seat in the kitchen, I was able to make the transition to a more sustainable lifestyle.

Through my experience, I learnt that it is important to acknowledge that not everyone will relate to certain issues the way you do. I found success in reducing my kitchen waste through collaborating with the people I live with, by converging on interests like health, zero waste and mindfulness.'

ZERO—WASTE RECIPES

Here are three south Indian dishes that use completely natural ingredients, which is obviously great for your health. Yet, these ideas that come from Karnataka, Kerala and Tamil Nadu provide more than that. They promote preparation and cooking that uses 'root to stem' methods, where all of the ingredients are used.

They are recipes that promote a waste-free life. Try these at home!

Mama Mansoor's Granola Recipe

You will need:

- 4 cups of old-fashioned oats,
- 2 cups of sliced almonds,
- 1 cup of dried cranberries (optional),
- 1 cup of dark raisins,
- ½ cup of chopped walnuts,
- ¼ cup of brown sugar,
- ½ cup of honey,
- ½ cup of vegetable oil,
- ½ teaspoon of vanilla extract,
- ½ teaspoon of ground cinnamon,
- ¼ teaspoon of salt.

Try to source all the ingredients from a market or whole foods store where you can take your own containers and buy the quantity you want.

Follow these straightforward instructions for this all-natural breakfast feast:

- Preheat the oven to 150°C,

< 78 >

- In a large bowl, mix the oats, almonds, brown sugar, salt, walnuts and cinnamon. Set aside,
- In a large measuring cup or bowl, whisk together the honey, vanilla and vegetable oil. Pour over the dry ingredients and mix everything well,
- Spread the mixture evenly in a 10 x 15-inch baking pan,
- Bake for 35–40 minutes making sure to give it a toss every 10 minutes for even cooking,
- Once it comes out of the oven, toss in the raisins and cranberries,
- Seal in an airtight container and enjoy at leisure.

Ridge Gourd Peel Chutney (Peerkangai Thol Thogayal)

(You can substitute ridge gourd peels with chayote/choko peels, or even snake gourd seeds, along with the white core).

You will need:

- 3 ridge gourds,
- 2 to 3 red chillies,
- 2 tablespoon urad dal,
- ½ teaspoon mustard seeds,
- ¼ teaspoon asafoetida,
- 10 gram tamarind (roughly the size of a small lime),
- Salt to taste,
- 2 teaspoons oil.

This recipe, too, ensures that you use all of the products. It is a healthy, zero-waste way to utilize the peels that are often left uneaten.

For a truly tasty chutney, follow these steps:

- Peel the ridge gourds and save the peels after removing the ridges,
- Wash the peels well and keep aside,
- Heat oil in a pan and add the mustard seeds and asafoetida,
- Once the mustard seeds splutter, add red chillies and urad dal. Stir until the dal turns golden brown,
- Add the peels to the pan and keep stirring for 3–4 minutes until they are cooked,
- Take the pan off the flame and add the tamarind. Stir while the pan is still hot.
- Once it cools down to room temperature, grind the peel and spice mixture, along with the salt, into a chutney,
- Serve it as an accompaniment with steamed rice, rotis, or as a spread for breads.

Nendran Banana Peel Curry (Kayatholi Thoran)

Nendran banana is native to Kerala and is used for making the very popular chips that look like raw plantain chips but are yellow in colour.

You will need:

- 250 gram Nendran banana peels,
- 3 tablespoon freshly grated coconut,
- 1 teaspoon cumin seeds,
- 2 red chillies,
- ¾ teaspoon mustard seeds,
- A few sprigs of curry leaves,
- 2 teaspoon coconut oil (optional),
- Salt to taste.

< 80 >

This recipe involves using the peels of this variety of banana. If you're reading this recipe in India, you can help save waste from local stores that stock these chips! Many of these snack shops simply discard the peel. If you don't have access to the fresh fruit, you can procure the peels this way.

To prepare this dish, all you need to do is:

- Soak the peels for 10–15 minutes,
- Wash them well and chop the peels into small pieces,
- Remove black spots, if any, while chopping,
- Bring a pot of water to boil and add the chopped peels to it,
- Add salt and cook for about 5 minutes until the peels are cooked,
- Strain and keep aside,
- Grind coconut, cumin seeds and chillies into a coarse paste,
- Heat oil in a pan and temper the mustard seeds, urad dal and curry leaves,
- Once the seeds splutter and the urad dal turns golden, add the ground coconut, cumin and chilli paste,
- Stir for 2–3 minutes,
- Add the Nendran banana peels and some coconut oil for some added flavour (optional),
- Cook on a low flame for 10 minutes,
- Serve hot with rice and dal, or rotis.

Note from a cook: Although the three recipes above have been described as techniques from Karnataka, Tamil Nadu and Kerala, there are many commonalities among all the southern states of India. This means that these recipes may appear to be similar to others from neighbouring locations (the same is true of kitchens all over the world, food and good ideas are shared from plate to plate, region to region).

For instance, the classic Tamil cookbook *Samaithu Paar*,[4] which translates to 'cook and see' has recipes that use Nendran banana and plantain peels to make curries. Highlighting the importance of good health and tasty food, and ensuring that ingredients are not wasted, have been relevant throughout history. It's really important to share ideas, which, with any luck, will help kitchens across the globe move to less wasteful practices by using up all parts of an ingredient for yummy breakfasts, lunches and dinners, or those healthy snacks!

'As a country, we are in an amazing position today where we can see the way things were in the past with our vibrant bazaars and markets. This lets us assess how things have changed in the big cities and towns with the advent of food-delivery companies and branded supermarket chains that sell only pre-dictated amounts of packaged food. Keep in mind that when packaged items began to appear, they were really helpful in getting women into the workforce. However, now that we know the harmful impacts, we have an opportunity to develop new solutions that support both social and environmental objectives.

The steps need to involve all stakeholders from across the country, from the farmers and their cultivation practices to our plates and the food we eat to the composted nutrient-rich soil at the base of the tree. By understanding how things work, we can once again begin to feel more attached with nature and the food we eat.

With a little luck, it might mean that you will live a happier, healthier life because you are more mindful of how profoundly interconnected the entire food system is when it comes to reducing waste, good health and overall well-being.'

ZERO—WASTE TIPS AND TRICKS

- Compost your organic waste,
- Bring reusable bags to shops and buy fruit and vegetables that are package-free (arm yourself with a reusable water bottle, a couple of grocery totes, a few cloth bags, reusable jars and bottles),
- Buy milk and eggs in reusable containers,
- Reuse and reinvent your leftovers,
- Buy local and in bulk,
- Use root-to-stem methods for all of your cooking,
- Stop using non-stick pans. Find alternatives, e.g., steel,
- Use reusable containers for your ingredients,
- Make your own products instead of buying takeaway,
- Use reusable cloths and organic scrubbers to clean,
- Invest in a pressure cooker (cuts the cooking time by half),
- Welcome alternatives to disposables such as paper towels, dustbin liners, baking wax paper, aluminium, plastic cling film, disposable plates, cups, etc.,
- Swap paper towels for reusable rags,
- Swap sandwich bags for kitchen towels or stainless containers,
- Drop plastic bin liners all together (as wet waste is compostable).

< 83 >

KITCHEN - EXTRACTION TO PROCESSING

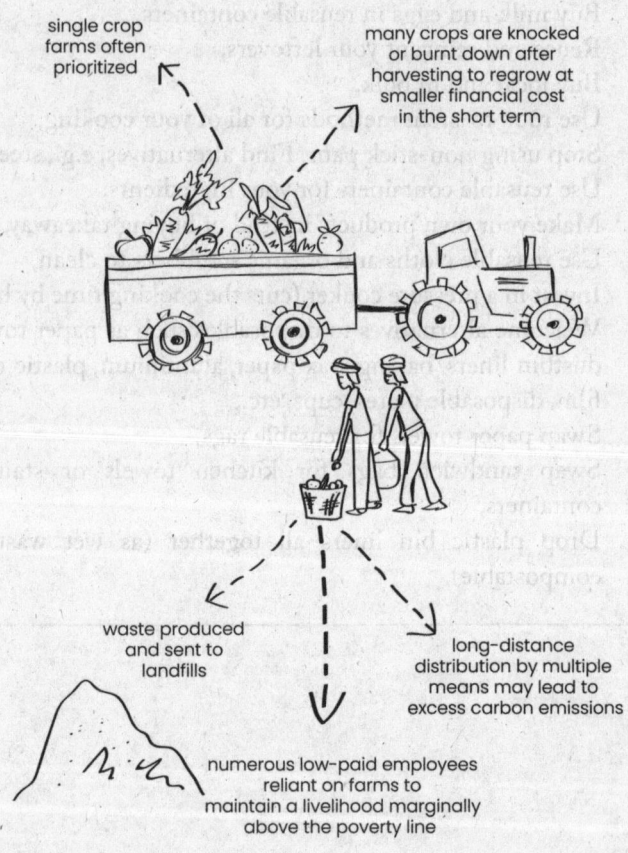

single crop farms often prioritized

many crops are knocked or burnt down after harvesting to regrow at smaller financial cost in the short term

waste produced and sent to landfills

long-distance distribution by multiple means may lead to excess carbon emissions

numerous low-paid employees reliant on farms to maintain a livelihood marginally above the poverty line

KITCHEN - CREATION TO LOCAL PRODUCTION

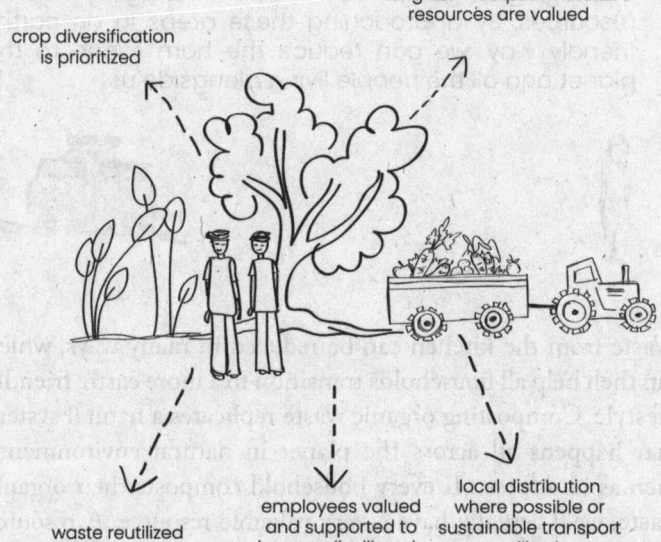

crop diversification
is prioritized

crops managed for
long-term output where all
resources are valued

waste reutilized
as compost thus becoming
a resource

employees valued
and supported to
improve livelihoods

local distribution
where possible or
sustainable transport
utilized

Why Is It Harmful?

> 'I find the kitchen a truly fascinating area to think about. There is a huge number of interconnected steps from the time the item is produced until well after it is on the chopping board. Learning why kitchen items (organic and non-organic) produce waste will really help you see how to start incorporating zero-waste practices into this aspect of your life. In this section, we'll focus on the impacts of product packaging, organic waste and managing natural resources. By approaching these areas in an earth-friendly way, we can reduce the harm done to the planet and all the people living alongside us.'

Waste from the kitchen can be reduced in many ways, which can then help all households transition to a more earth-friendly lifestyle. Composting organic waste replicates a natural system that happens all across the planet in natural environments, such as in a forest. If every household composts their organic waste, we'll actually have a very valuable resource. A resource that is able to nourish the soil and enable new living things to grow, including trees, vegetables and fruits. There are many positive steps just like this one that can benefit both humans and the planet.

Impacts of Product Packaging

'Cling film is difficult to recycle. The majority of it ends up in landfills where it takes hundreds of years to degrade and risks leaching chemicals into groundwater.'[5]

< 86 >

Food wraps, such as cling film, gained popularity during the 1950s in part due to poor food-storage systems.[6] The materials increased in effectiveness and type in the decades that followed and have, consequently, led to plastic infiltrating the environment. When the wraps end up in marine environments, they harm the marine animals who mistakenly consume it as food.[7] Part of this pollution is due to the fact that recycling the thin material is more costly for a business than using virgin materials[8] to create new products,[9] which means that the material people thought they recycled actually ends up in landfills and the ocean.

This is just one example. Others that produce similar plastic waste include the use of single-use containers such as drink bottles. In fact, a 2019 report highlighted that a major soft drink company is the most polluting brand worldwide. The company was found to be responsible for 11,732 pieces of waste found in thirty-seven countries across four continents.[10] Similarly the use of single-use plastic bags that are typically used for less than twelve minutes, kills, on an average, 100,000 marine animals, while only 1 per cent of all bags produced is recycled. Also, the material can take in excess of 500 years to break down,[11] giving us a powerful illustration of the environmental effects of buying groceries on a given day.

The world is waking up to these issues. Many countries have begun to implement bans on these items,[12] yet the waste that they produce will be determined by the practices that you as a consumer undertake while shopping and how you store your products at home. A move to a circular system, where you use reusable bottles and bags, is a positive step in eliminating this problem worldwide.

Organic Waste

'It is estimated that saving one-fourth of the food currently lost or wasted globally would be enough to feed 870 million hungry people in the world.'[13]

The Food and Agriculture Organization (FAO) reports that a third of food waste produced each year is wasted or lost (approximately 1.3 billion tonnes).[14] To tackle the problem, numerous stakeholders have aimed to cut down food waste by half by 2030, as per criteria suggested in the Sustainable Development Goals.[15]

Stakeholders with different degrees of influence are assisting in achieving these goals. They include the government of France, which passed a law that prevents any food waste from being thrown into landfills.[16] On a smaller scale, the social business named Soul Kitchen, an Italian initiative, 'demonstrates transformative examples of equitable and inclusive programs designed to promote a just food system and advocate for the value of human potential',[17] among other things. A large part of their primary focus is the necessity to understand the entire supply chain and utilize available resources. Similarly, the Real Junk Food Project in the UK uses food thrown away by homes and businesses to feed people who cannot afford it.[18] This has, in turn, saved a substantial amount of waste ending up in landfills, which would otherwise have contributed to climate change via methane production.[19]

Progress has been made, but more can and needs to be done, especially in countries such as India with large populations but, to date, limited tangible implementation of food-saving systems at the national level.[20]

'The major challenge for many developing countries like India is in the process that the food undergoes before it reaches the end consumer. Although food wastage is a global problem, India stands a chance to convert this into an opportunity . . . The world's second most populous country needs to reduce its food wastage to feed the 194 million Indians who go hungry daily. It is important that technology is adopted at every stage of the supply chain to overcome this problem. The innovative reefer-container technology has reformed the transportation of perishable goods, helping food remain fresh for over a month, enabling agriculture producers to safely send all products ranging from grapes to shrimps across geographies, efficiently.'[21]

New methods within a supply chain, or the changing habits of consumers to consume only in-season produce,[22] will limit the amount of waste created throughout the overall process. Such targets are essential if there is to be a greenhouse gas reduction to required levels in order to limit climate change,[23] as well as to create a sustainable food system.[24, 25] The interconnected nature of all systems on the planet is becoming clearer to a growing number of stakeholders, such as Zomato Feeding India and Robin Hood Army that work across India and Pakistan. These organizations attempt to 'connect hunger and food waste as solutions for each other'.[26] They aim to feed the impoverished rather than letting the food go to waste, collecting unwanted food from a range of sources, including individual families and from events such as weddings.[27] Initiatives in South Asia and many other locations around the world, run by local organizations or implemented by governments, have the potential of providing a range of positive outcomes.

Managing Natural Resources

'In India, approximately 80 per cent of farmers are poor, marginal producers. Among the challenges they face are fluctuating market prices. A tomato can sell for US $0.28 [Rs 20 approximately] one month and only US $0.03 [Rs 2 approximately] one month later. However, farmers are not able to preserve their food for more than 1 to 2 weeks because of electricity costs, poor infrastructure and lack of funding to invest in storage facilities.'[28]

One area causing problems is growing crops that are not eco-conscious. Globally, there is a reliance on large yields of single crops (monoculture). Not only does this not allow farmers to counter external shocks like droughts and floods,[29] it also damages the soil, increases salinity and raises the probability of diseases in plants.[30, 31]

This scenario can adversely affect livelihoods because farmers (especially in lower-income countries) are regularly the lowest earners, particularly small-holder farmers.[32, 33] Actions away from this system can include using perennial crops such as kernza, a domesticated wheatgrass that has many applications. This can be seen as a key move towards sustainable agriculture. Perennial crops can help farmers achieve a higher livelihood and allow the environment to regenerate, which could help in achieving levels of biodiversity that are beneficial for the planet.[34]

Furthermore, an area of untapped potential to improve the sustainability of farming is through carbon capture, farm fields—not just forests—have enormous carbon storage potential. Maximizing the potential for carbon sequestration on our croplands has potential benefits for myriad stakeholders, from farmers to policymakers, business leaders, investors and consumers. This would have the dual benefit of increasing soil carbon content while also producing more profitable and resilient food systems that are better capable of withstanding the effects of climate change, such as the increased severity of droughts, pests and diseases. Also, farmers will get healthier fields; taxpayers get a healthier planet—a clear win-win.[35]

Within the food system, there is a need to share more information about how to improve it. This can be helped by creating a more transparent outline of what is occurring throughout the process. This will not only benefit the environment but also the workers in the system. It can help reduce instances of child labour in the production of chocolate in countries such as Ghana and Côte d'Ivoire.[36, 37] It can also provide roadmaps going forward, such as reducing post-consumed discards, which has, to an extent, been successfully achieved in South Korea.

South Korea, through 'smart bins'[38] and taxes, along with raising awareness of the issues concerning food waste, has significantly improved the amount of organic products that are composted. Statistically speaking, the country had only utilized 2 per cent of organic waste in 1995, while in 2019, due to the initiatives put in place, 95 per cent of organic waste was a resource.[39]

Changes to the food system made in South Korea, and encouraged by the Sustainable Development Goals, are sound templates that need to be implemented worldwide. Yet, this issue is not a 'one size fits all'. There is a need to find solutions that suit each location, whether that is composting through farming cockroaches in China,[40] or looking into new technologies and ways to do farming that limit the effect on the environment, such as vertical farming.[41]

Further illustrating the fact that the same solutions may not work everywhere, other countries have found their own solutions. In Ethiopia, for example, where there is a huge problem of mixed waste in the capital Addis Ababa, a waste-to-energy plant has been set up. It reduces the amount of waste in landfills without emitting toxins into the air due to the filtration systems in place. This, in turn, allows the inhabitants in the growing capital city to breathe cleaner air, thereby improving livelihoods.[42, 43] In the state of Chhattisgarh in India, the government has developed multi-commodity cold storage chambers and implemented post-harvest management programmes to 'establish a network of pack houses and cold storage, to help prevent the damage of perishable commodities including fruits, vegetables and flowers, in the entire state'.[44] All of these methods use innovative technology and systems that suit the area, with the ultimate aim to reduce waste.

Compared to these uses of technology, the World Wildlife Foundation (WWF) has highlighted simple steps to reduce food waste[45, 46, 47, 48] that do not require advanced technology. When food is wasted, all 'energy and water it takes to grow, harvest, transport, and package it'[49] is expended.[50] Additionally, when 'food goes to the landfill and rots, it produces methane—a greenhouse gas even more potent than carbon dioxide. About 11 per cent of all the greenhouse gas emissions that come from the food system could be reduced'[51] if humans stopped wasting food.

Worldwide, there has been a huge loss of food potential in the environment through overfishing,[52] deforestation[53] and droughts.[54] Acting more sustainably and thinking of the circular system, with set metrics about how much food can be taken based on how much is consumed,[55] along with a transparent supply chain throughout the process, is paramount to success.

A circular system can also lead to damaged environments re-establishing themselves. Sustainably managing the land in a waste-free fashion, which replicates methods found in the natural environment, while utilizing modern technology can have a positive impact on people and the planet.[56, 57]

What Can You See in Your Waste Now?

'I hope you're enjoying the lessons that you are learning. I always find it profound to reread how one action impacts the others. It really inspires me to keep trying to make a difference.

There are a few more (super yum) recipes at the end of this chapter, after this final activity. Our next step after this will see us look at our homes as a whole, so I'll meet you there!'

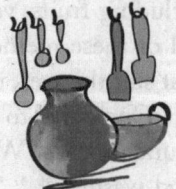

LOOK A LITTLE DEEPER WITH THIS ACTIVITY!

Activity #3 is a four-part process, on separate question sheets, that builds on everything you have learnt about this topic throughout the chapter. The importance of this exercise is to understand and become aware of the processes that can help you transition to a sustainable lifestyle.

ASSESS YOUR
KITCHEN WASTE:
ACTIVITY #3 QUESTION 1 OF 4

What type of environmental impact do your products/services have?
Join the product/service to the problem using an arrow:

Product/Service:

Vegetables

Storage containers

Pulses

Problem:

Chemical pesticides

Wasteful watering methods

Plastic pollution

Use the ideas from this sample to assess your waste. The product/service may relate to more than one problem. Fill in the lines below with your products/services and environmental problems.

_____   _____

_____   _____

_____   _____

_____   _____

_____   _____

When thinking about the environmental impact, it is important to think of the number of areas that it could affect. Start on a small scale and work your way out. First, think about what it means to the micro-environment around you, then think larger and larger until you look at it from a global perspective. It will be beneficial to undertake some research online or in a library.

ASSESS YOUR
KITCHEN WASTE:

What system issues prevent change?
Join the product/service to the issue using an arrow:

Product/Service:	Issue:
Vegetables	Government regulations
Storage containers	No water limits enforced
Pulses	No affordable alternatives

Use the ideas from this sample to assess your waste. The product/service may relate to more than one issue. Fill in the lines below with your products/services and environmental issues.

_____   _____

_____   _____

_____   _____

_____   _____

_____   _____

A system is anything associated with the product or service that is interconnected with it, for example, the manufacturing unit that produces the product or the government office that provides the service. To learn more, conduct research online or in a library.

ASSESS YOUR
KITCHEN WASTE:

What sustainable options are there to replace it?
Join the product/service to the solution using an arrow:

Product/Service:

Vegetables

Storage containers

Pulses

Solution:

Support chemical-free farmers

Home gardens

Invest in steel or ceramic

Use the ideas from this sample to assess your waste. The product/service may relate to more than one solution. Fill in the lines below with your products/services and solutions.

———————   ———————

———————   ———————

———————   ———————

———————   ———————

———————   ———————

A sustainable option is a product or service that will last longer and/or produce less waste. Think about options such as products made from earth-friendly materials, or those that reduce waste through a supply chain. To learn more, research online or in a library.

ASSESS YOUR
KITCHEN WASTE:

How will you start using the sustainable option?
Join the product/service to the action using an arrow:

Product/Service:	Action:
Vegetables	Support industry change
Storage containers	Grow your own
Pulses	Use containers that last a lifetime

Use the ideas from this sample to assess your waste. The product/service may relate to more than one action. Fill in the lines below with your products/services and actions.

_____   _____

_____   _____

_____   _____

_____   _____

This last step is all up to you. Make your choice, know the benefits and live a zero-waste lifestyle. You are more likely to succeed with support from your friends and family.

ZERO—WASTE LIBRARY

Copper + Clove's Broccoli Almond and Feta Cheese Root-to-Stem Salad[58]

This recipe can be made in the quantities that you need, based on how many people you are serving. It's really tasty and uses the ingredients from root to stem, aiding in good nutrition and zero-waste practices.

You will need:

- Broccoli,
- Almonds,
- Feta cheese,
- Cherry tomatoes,
- Lime,
- Salt,
- Mint leaves,
- Coriander seeds,
- Pepper,
- Groundnut oil.

Now that you have gathered all your fresh, healthy ingredients, the next steps are as follows:

- Cut the broccoli into little mini trees (keep the leaves),
- Cut the stem into small pieces (full of fibre and other nutrients). The end of the stem that's exposed can be composted,
- Add a little bit of lemon juice, lemon zest, salt and pepper to the broccoli and stir-fry until it is a little golden brown (in an oil of your choice),

- Cut the almonds, feta cheese and tomatoes into same-sized pieces,
- Take the chopped almonds and lightly toast them,
- Add some coriander seeds that have been lightly toasted,
- The dressing is made with lemon and oil. Zest the lemon and then chop into pieces.
- Also squeeze your lemon into a container (you will need about 3 tablespoons),
- Take the cherry tomatoes and put them in the oven for 20 minutes with a little salt and oil.

To assemble your super-yum salad, follow this method:

- Add broccoli into a big serving bowl,
- Add tomatoes with all the roasted juices too!
- Add feta cheese, marinated in herbs and oil, with paprika powder,
- Add mint leaves,
- Scatter the toasted almonds and coriander seeds,
- Add half a tablespoon of dressing—groundnut oil, or you can use any other oil you of your choice,
- Add half a tablespoon of lemon juice,
- Season with salt and pepper,
- Sprinkle microgreens on top.

Thanks are due to Sarah Edwards from Copper + Cloves in Bengaluru for providing this amazing root-to-stem recipe!

The Reset
This is a detox drink.

You will need:

- 1 tablespoon lemon juice,

- 1 tablespoon apple cider vinegar,
- 1 teaspoon honey,
- 1 teaspoon grated ginger,
- 1 cup of warm water,
- An optional pinch of pepper, or two whole dried pepper balls.

Get your cup and get ready to reset:

- Pour some warm water into your cup.
- Pour all of the ingredients into the cup, in no particular order.
- Stir and drink up!

Peanut Butter Espresso Munch Smoothie Bowl

You will need:

- 1 generous tablespoon of peanut butter,
- 1 chopped banana,
- 1 or 2 pitted dates,
- 1 cup coconut milk (or almond milk or oats milk),
- 1 espresso shot/cup of coffee,
- A pinch of cinnamon.

The optional ingredients include:

- 2 or 3 ice cubes,
- 1–2 tablespoon of hemp protein,
- 1–2 tablespoons of pumpkin seeds, chia seeds, halim seeds (or garden cress seeds), flaxseeds and granola.

You're probably going to want to take a picture of this when you're done. Follow the method to have some fun:

- Add all the ingredients into a mixer,
- Blend together until it has a smoothie-like consistency,
- Pour it into a bowl,
- Sprinkle pumpkin seeks, chia seeds, halim seeds, flaxseeds and/or granola on top.

Your perfect post-workout smoothie bowl is ready!

Nature's Pre-Workout Energy Balls

You will need:

- 1 cup of pitted dates,
- ¼ cup almonds, chopped. Dry-roast them,
- ¼ cup cashews, chopped. Dry-roast them,
- ¼ cup grated dried coconut.

The optional ingredients include:

- 2 tablespoons of fried goondh (dink) in ghee,
- Flaxseed and halim seeds.

This one is just what you need for a busy, busy day. It is inspired by Copper + Cloves and Mama Mansoor. Here is the method:

- Add the dates, almonds, cashews and coconut to a mixer,
- Pulse twice or thrice until the contents are combined,
- Add in the fried goondh (dink),
- Once it is all fluffed up, crush them,
- Take a mixture of flaxseed and halim seeds and roll the paste over into laddoos.

Dal Moong More

This is a moong dal dosa recipe.

You will need:

- 1/2 cup split green moong dal,
- Grated vegetables (your choice, carrots are fantastic with this),
- A finely chopped onion,
- 2 green chillies,
- 1 teaspoon grated ginger,
- A pinch of jeera powder (cumin) and salt to taste.

All you need to do is:

- Wash and soak the moong dal for 2 hours and grind it to a paste,
- Add grated vegetables,
- Chop the onion and green chillies for the garnish.

You can make chilas or dosas and serve them with chutney or mixed peanut butter and tamarind chutney.

Zero Wa-Stew

This is a comforting vegetable stew.

You will need:

- Vegetable scraps,
- Water,
- Salt,

This one is a big waste saver. Use any leftover vegetables that you have.

< 102 >

Ready for the recipe? It's bound to be a hit.

- Throw all your food scraps (onion bits, broccoli stems, tomato tops, anything and everything, really) into a large pot,
- Add water so that it covers the scraps,
- Let it simmer for about half an hour,
- Add salt,
- Strain the scraps,
- Set a metal strain inside a metal bowl and dump everything into the strainer,
- Lift out the colander and reserve the scraps for the compost pile,
- Pour the broth in jars (keep a strainer handy, one that doesn't catch the superfine particles).

You will love using the broth for the base of your next meal!

Zero-Waste for Zero-Hunger Local Organizations

Listed below are a couple of local organizations focused on reducing food waste and feeding people who often miss out on daily meals.

- Zomato Feeding India works in more than 100 cities in India. Their aim is to work towards 'better food for more people' and aiming to ensure that there is 'food for everyone', with their ultimate goal being to end hunger across the country.[59]
- Robin Hood Army 'is a volunteer-based zero-funds organization that works to get surplus food from restaurants and the community to serve the less fortunate people'.[60] They work in India and Pakistan with an ultimate goal to 'beat hunger and bring out the best of humanity using food

as a medium. The idea is to create self-sustained chapters across the world that will look after their local community. And in the process, inspire people around us to give back to those who need it most.'[61]

You can do your own research to find out if there is an orphanage, school, elderly home or migrant community in your neighbourhood that may benefit from a donation.

Community-Based Composting and Biogas Solutions[62]

Most of the businesses listed below, you will note, are Bengaluru-based. Fortunately, many of them ship across the country, have partners in other locations, or can even distribute waste solutions internationally.

- Daily Dump, located in Bengaluru,
- Shudh Labh Solutions Pvt. Ltd, located in Bengaluru,
- Pelican Biotech, located in Kerala, Karnataka and Tamil Nadu,
- One Hop Organics Pvt. Ltd, located in Bengaluru,
- Prudent Ecosystems, located in Bengaluru,
- Stonesoup, located in Bengaluru,
- Soil and Health Solution, located in Bengaluru,
- Shree Skanda Solar Systems, located in Bengaluru,
- Spinform Plastics Pvt. Ltd, located in Aurangabad,
- GreenOn, located in Bengaluru,
- Vennar Organic, located in Bengaluru,
- GPS Renewables Pvt. Ltd, located in Bengaluru,
- Vermigold Ecotech Pvt. Ltd, located in Mumbai,
- E Save Promoters, located in Bengaluru,
- Eco Positive Solutions, located in Bengaluru,

- Quantum Green, located in Bengaluru,
- Quantum Leaf, located in Bengaluru,

You can learn more about composting and biogas on many of these websites.

make your
own surface cleaner

hang dry clothes
on a line instead of
using a dryer

use natural cleaning
alternatives

ZERO-WASTE
HOME
CARE

buy organic
detergents in
bulk

plan laundry days
to save water
and energy

place house plants
in rooms to help
freshen your home

use reusable rags,
cloths and mops
to clean

CHAPTER 4

HOME CARE

'Harmful components of detergents such as anthropogenic components . . . Herbicides, pesticides and heavy metal concentrations (e.g., zinc, cadmium and lead) can cause the water to grow murky, thus blocking out light and disrupting the growth of plants.'[1]

The Perfect Life . . .?

You had a party yesterday and your home is a mess. You gather the table covers and take them to your laundry room. You put them in the washing machine with a single-use packet of detergent and hit 'Start'. The machine chugs along as you gather other cleaning supplies. The surface cleaner came at a great two-for-one deal. You rip open its plastic packaging, tear out a paper towel and get to the cleaning.

Once done, you pick up your feather duster, wondering why it is even called that because it is made of plastic. Your broom sweeps up the dust and you mop the floor with a cleaner that has a strong lemon scent.

You finish up after a couple of hours and place your sheets in the dryer. They spin around in the machine as you walk around the house that is aglow with all the lights you left on. You leave your house with the fans and lights switched on, thinking that perhaps they will help the floor dry faster. Your electricity bill was reduced recently anyway, thanks to a community tax that was waived off.

Your friends were talking about different forms of energy last night. Things you had thought of in passing but never in detail, especially solar and wind energy. You remember laughing a little because of the cheaper deal that you got thanks to your community; you don't need to worry about a thing.

You hop into your car and make your way to the local store for a few items that were knocked over the night before. Nothing expensive, completely replaceable.

You notice a magazine while you are there that speaks about sustainable architecture, limiting waste and greening up your home. It seems a little ridiculous to you. You have your gardener to manage the couple of plants out front and you see no reason why your housing complex needs to be greener.

You come back home and place the new items where they're supposed to be and take the broken items to the roof of the multi-storeyed complex you live in, where other such broken items are stored as well. It is hot outside and the pile is larger than you remember. You make a mental note to mention this problem to the landlord so that he can get these things dumped in the landfill.

You look across to the other roofs and notice that they don't have the solar panels that your friends mentioned, nor do they look particularly green with the 'green roof' concept the magazine had a photo of. Strange times, you think, as you walk downstairs to your favourite seat, turn on the television and let the fan cool you off as you hear the beep of your dryer in the background.

What Is in Your Waste?

'Many laundry detergents contain approximately 35 per cent to 75 per cent phosphate salts. Phosphates can cause a variety of water pollution problems.'[2]

'I would like to start off with a question for all of us to think about. Are you attached to the products you clean your home with?

During my assessments of my own waste practices, I found this to be quite a fascinating question. For cleaning, many people in India rely on domestic help. We have also followed modern trends of installing robots, such as washing machines, in our homes that separate ourselves from the clothes that we wash. This could never have been dreamt of while families hand washed clothes in rivers and streams.

In order to reduce all the forms of waste that we produce, we really need to become aware of this detachment from old-school techniques and the attachment to new technology. Don't get me wrong! I totally acknowledge that these innovations or robots have allowed us to save time and get out of our homes and pre-dictated domestic roles to pursue our passions and careers. However, through this attachment, we have begun spiralling towards this delusion that we don't have time to slow down, thereby chasing convenience. This has continued adding to the phenomena of mass production to meet intense demand, planned obsolescence and mounting e-waste. By simply slowing down and finding the balance between technology and these amazing old-school techniques, I came to realize that we actually don't need to be as obsessed as we are when it comes to buying new robots, products or even home cleaners. This made me feel liberated from all the marketing pressure from various corporations that simply seek profit. It really helped me introspect on my relationship with my home, and I started to adopt a more anti-consumerist approach.

I have found that if we are to become aware of the environmental impact we have, we must become more knowledgeable about the products we use. It is thought-provoking to ask how we position ourselves in our homes. Do you see yourself and your home as a part of a bigger ecosystem, or do you consider it to be completely independent of the broader world?'

< 110 >

NOW IT'S YOUR TURN TO ASSESS!

Try out Activity #1 with your current knowledge. View this activity as an initial way to see how much you know. Once you learn more as we go forward in this chapter, you can return to it in order to see how much your knowledge has developed.

ASSESS YOUR
HOME CARE WASTE:
ACTIVITY #1

Follow the example sheet below by choosing a few
of your own products to assess.

	Example A	Example B	Example C
Does the product have ingredients that can harm the environment?			
Is the cleaning product you are using natural?			
Are there more sustainable solutions to the current methods that you are using?			
Do you know all the consequences of your home-care practices?			

- If yes: share the solutions with friends and family.
- If no: research options.

Use the ideas from this sample to assess your waste. You can draw your own sheet based on
this and create other questions for this assessment based on your needs.

What Resources Are Available?

'I've never been someone who focuses on home care, but after I addressed my personal-care products, it seemed a natural progression to start thinking about the rest of my home.

During my research, I found out that the major sources of pollution in some of Bengaluru's rivers and lakes were the chemicals used in detergents and home cleaners. This was a startling realization! As a result, I began to look at the list of chemicals at the back of each container.

An alarming example was when I found that there is a common misperception about the lather we experience when using soaps. Most of us associate the amount of lather with cleanliness and the quality of the cleaning agent. However, very little lather is required to trap the dirt on our dishes (or skin and hair for that matter)! Therefore, the excess lather is simply there to fool us consumers.[3] I found more information on the ingredients listed in letters and numbers, as well as the long words that I could hardly pronounce. It was a terrible awakening.

I became aware of the sales gimmicks in the linear process of production and consumption. I started noticing the campaigns that work in a sensationally successful way to fill their already cash-full pockets, horrendously damaging the environment in the process.

I began to realize that despite the fact that homes are such an important part of Indian culture, the detachment that many of us feel to the items we use is causing wasteful practices. It means that, often, we do not understand the implications of what we are buying. Instead, we focus on two things: what it costs us and the product's perceived value.

In terms of cost, we are generally willing to spend less on home-care products than personal care because it is not explicitly in contact with our bodies. We think it will have no effect on our health. We often justify our reasons for buying products through commercials that highlight false perfection and product packaging. These factors can lead to increased use of harmful chemical ingredients that are one of the causes contributing to the frothing of rivers and lakes in our cities.

One of the solutions I now use is thanks to Vani Murthy. She is an advocate for waste reduction, a renowned urban farmer and a passionate citizen leader! She hosts her own YouTube channel,[4] among other teaching outlets such as Instagram. I followed her lessons, learnt about her bio-enzyme recipe and how to make laundry detergent. She is a phenomenal mentor. Through her lessons, I found a natural way of cleaning. Some of my favourite home-care recipes are in the next section. I'll meet you there after this activity.'

TRY THIS ACTIVITY OUT!

Throughout the chapter there are a number of suggestions that you can use for Activity #2. You can read ahead and return with new knowledge, or assess yourself now before learning a little more and returning to this exercise again.

< 114 >

ASSESS YOUR
HOME CARE RESOURCES:
ACTIVITY #2

Find one to three products that you can make yourself and/or add some indoor plants to your home to improve air quality. Use the suggestions in the guide book or research to find more. Fill in each section of this activity to record your achievements.

The resources that I have learnt about are:

 A quick and easy vinegar and sodium bicarbonate (baking soda) surface cleaner.

 Soap-er Nuts multipurpose laundry cleaner.

 I added a peace lily and areca palm to my home.

Draw a picture from your activity to show friends and family:

The resources needed for this activity are:

 White vinegar, sodium bicarbonate or baking soda, lemon.

 Water, soap nuts.

 I visited a nearby garden nursery to pick the best plant for my home.

Record where you learnt about these resources:

 The guide book's 'Zero-Waste Recipes', 'Zero-Waste Library' and 'Why Is It Harmful?' sections.

Record who you shared your success with:

 My partner! We cleaned our entire house and added the plants together.

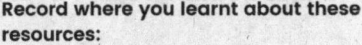

Use the above ideas to assess your resources and then create your own activity sheet.

Moving Towards a More Sustainable Lifestyle

'Many of the recipes I learnt from Vani and other environmental champions can be used for multiple surfaces. The ingredients will allow you to clean well while ensuring the process is earth-friendly. On top of that, if you make it in bulk at home it often works out cheaper. Here are two great recipes (there are more at the end of the chapter). Since the first one takes time to cure, I make it in bulk, while the second can be prepared in a much shorter time for spot cleaning.'

ZERO—WASTE RECIPES

Natural Multipurpose Cleaner

You will need:

- Jaggery,
- Orange peels (or any citrus peels),
- Water,
- Soap nuts,
- A jar or container that you can use while the ingredients develop into the finished product.

Follow these steps:

- Add ten parts of water into the jar,
- Add one part jaggery to this,
- Add three part orange peels,

< 116 >

- Stir well,
- Close the jar,
- Let it ferment/cure for three months. Make sure to note the date!
- Every morning, open it for a minute to release the gas (don't worry if you miss a day or two).

It will be ready in three months! The next stage will require a smaller jar for you to combine that mix of bio-enzyme with the remaining ingredients:

- Soak soap nuts in water,
- Once they are sufficiently saturated, add one part of the bio-enzyme to complete the process.

This mixture can be used as a natural multipurpose cleaner, or even a shampoo! It's best to store this in a reusable plastic jar (the fermentation may cause a glass jar to crack).

There are similar recipes in the 'Zero-Waste Library' section that do not require three months.

A Quick and Easy Vinegar and Sodium Bicarbonate Surface Cleaner

You will need:

- ¼ cup white vinegar,
- ¼ cup sodium bicarbonate (baking soda),
- Lemon,
- 2 cups water.

This is a quicker process compared to the previous recipe. All you need to do is:

- Pour vinegar into a container,
- Add sodium bicarbonate to the liquid,
- Do this slowly to manage the chemical reaction. This is worth the mention because of an experience I had where, in an attempt to multiply the batch, I increased the quantities by ten times. Instead of gently adding the sodium bicarbonate, I poured half a box into ten cups of vinegar, which caused a significant chemical reaction! The concoction sprayed freely over the entire kitchen, leaving me a little shocked and amused. I cleaned the kitchen up after that and have been cautious ever since,
- Squeeze in a lemon to lend it a citrus scent,
- Pour water in to finish the process.

'I found other sustainable methods, too, from traditional techniques. Like using coconut husks to clean the copper pots with my grandmother. I remember how the organic material worked really well with natural cleaners. Comparing this to the use of plastic sponges (which take hundreds of years to break down)[5] with cleaners that use carcinogens was extremely shocking.

While we're here, I want to share a cool recipe that I learnt from my grandmother. All south Indian homes have plenty of tamarind in their kitchens. Widely used for cooking, we add tamarind water to sambar, rasam and dishes like tamarind rice.

It turns out that besides being delicious, tamarind is also a handy cleaning agent. In fact, tamarind was used before dishwashing detergents were discovered. Tamarind is acidic, which is why it makes for a great cleaning agent, like vinegar and lemon. In fact, when it comes to cleaning, tamarind has an added advantage. It has a thick and hard crust that can be used to scrub oily surfaces.

Tamarind is mainly used for cleaning metals. In fact, the best way for cleaning silver, brass and other metals is by using this citrus fruit. If you don't want the pieces of tamarind to be scattered, you can use tamarind pulp. Also, you can soak your grubby copper vessels in tamarind water to soften the dirt that accumulates on them.

Here are some ways you can use tamarind for cleaning:

1. Kitchen sink: Can be scrubbed effectively with a piece of tamarind and some salt. It cleans all the water stains.
2. Silver: The metal becomes black when exposed to moisture or air. It can be returned to its former splendour with tamarind and salt.
3. Jewellery: Some intricate pieces of metal jewellery are difficult to clean with soap. You can soak them in tamarind water and then wipe them with a dry piece of cloth.
4. Brass: This is another metal that can sparkle if you clean it using tamarind. Old brass showpieces, clocks, and even door knobs, can be cleaned using tamarind pulp.
5. Copper vessels: Though these are rare these days, tamarind is perfect to clean them with.

During my transition towards a zero-waste lifestyle, I found that I needed to apply the natural process that I had introduced to my personal-care products to other areas of my home. I also found it beneficial to learn and share knowledge with the people around me. These processes helped in allowing us to lead a healthier life at home, benefit the environment, and, perhaps, change the habits of the people we interact with on a daily basis.

I'm sure you will find many of these things out for yourself once you try out some of the tips, recipes or study the illustrations, especially how new technology is incorporated. It's all about using the best bits of yesterday with the best earth-friendly methods that we develop today!'

< 119 >

ZERO—WASTE TIPS AND TRICKS[6] [7]

- Dry clothes on a line under the sun instead of in a dryer,
- Buy organic detergents in bulk,
- Plan laundry days to save water and energy,
- Use natural cleaning alternatives; you can make your own, e.g., surface cleaners,
- Use reusable cloths, rags and mops,
- Use alternative house cleaning tools like natural, compostable scrubbers such as bamboo,
- Place houseplants in rooms to help freshen up your home,
- Harvest rainwater or collect it to water your plants,
- Use chemical-free paints (low or no VOCs [volatile organic compounds]),
- Switch to energy-efficient appliances (LED, purchase energy- and water-efficient appliances),
- Install low-flow shower heads and toilets,
- Practice smart irrigation techniques,
- Install solar panels (if possible),
- Fix leaking toilets and faucets.

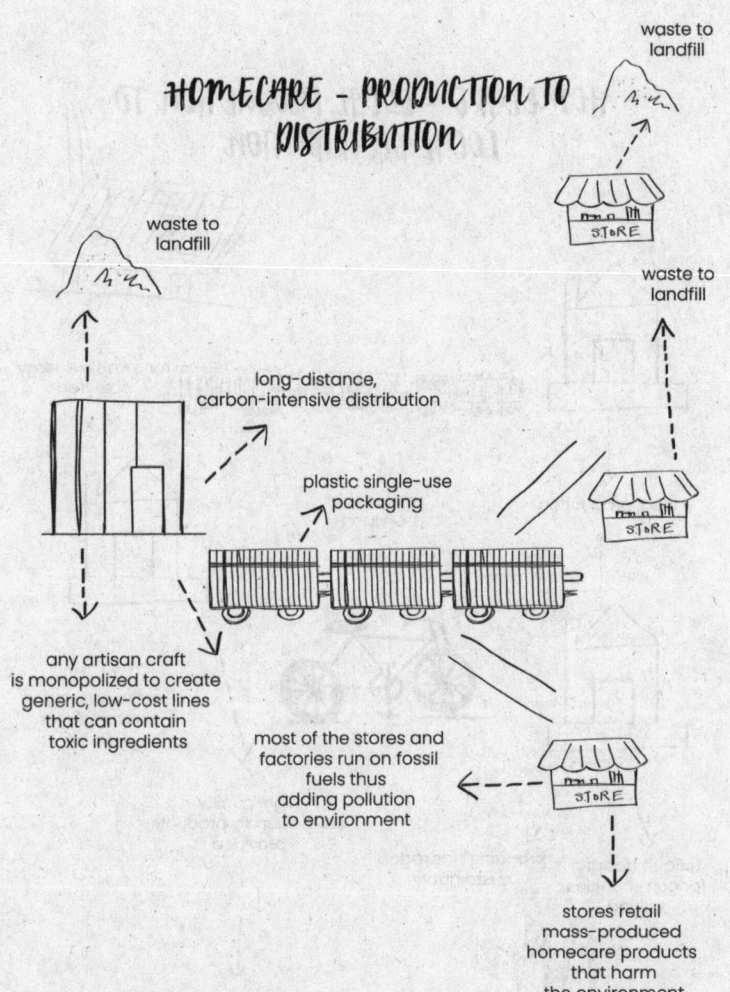

HOMECARE - PRODUCTION TO DISTRIBUTION

waste to landfill

waste to landfill

waste to landfill

long-distance, carbon-intensive distribution

plastic single-use packaging

STORE

STORE

STORE

any artisan craft is monopolized to create generic, low-cost lines that can contain toxic ingredients

most of the stores and factories run on fossil fuels thus adding pollution to environment

stores retail mass-produced homecare products that harm the environment

©Bare Necessities Zero Waste Solutions Pvt. Ltd.

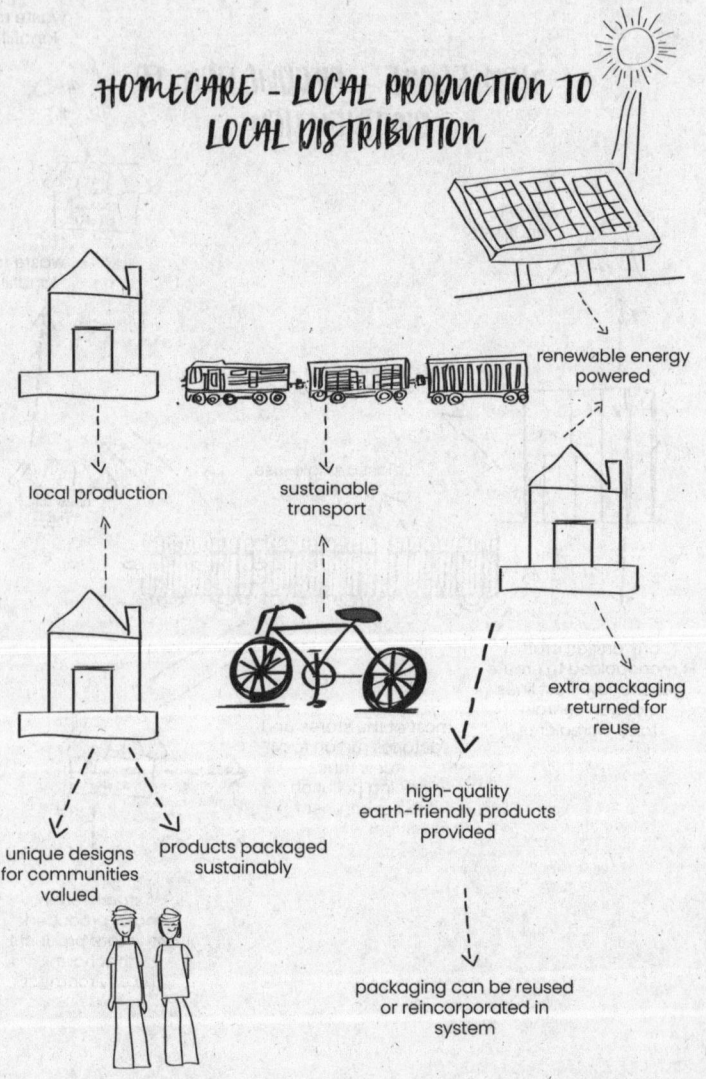

HOMECARE – LOCAL PRODUCTION TO LOCAL DISTRIBUTION

renewable energy powered

local production

sustainable transport

extra packaging returned for reuse

unique designs for communities valued

products packaged sustainably

high-quality earth-friendly products provided

packaging can be reused or reincorporated in system

©Bare Necessities Zero Waste Solutions Pvt. Ltd.

Why Is It Harmful?

'Some of the chemicals and terms that are discussed below may seem a little complex, but it is important to talk about them. The chemical additives in many products are worrying for our health and that of the planet. Only through learning about them can we make a difference. The same goes for all our sustainable practices. The more we know, the more educated decisions we can make, which in turn will help us produce less waste.'

The chemicals found in cleaning products can have harmful effects on your body and the environment. Examples include phthalates,[8] perchloroethylene,[9] triclosan,[10] quaternary ammonium compounds,[11] 2-Butoxyethanol,[12] ammonia,[13] chlorine[14] and Sodium Hydroxide.[15] These eight are considered to be among the worst offenders[16] to your body and the environment. Some of the impacts of these chemicals include reduction of fertility count in men, dizziness, loss of coordination, promotion of the growth of drug-resistant bacteria, skin irritation and kidney and liver failure.[17]

Chemical Additives

'We assume they are safe. But, in fact, many popular household cleaners are dangerously toxic.'[18]

Why are these products easily available?[19] A reason behind this is that in small quantities the chemicals are deemed to pose minimal risk. It isn't a reason that justifies banning the product according to the businesses manufacturing the items.[20] However, this measurement does not take into account the repeated effect they have throughout the entire system. It looks at it through a linear mindset where the true cost and damage that a product can cause throughout its entire lifecycle is ignored.[21]

Therefore, it's beneficial to evaluate products within the concept of a life cycle[22] and explore what this means at home and on a wider scale. You will similarly be able to utilize this understanding in other areas of your life, such as your closet or elsewhere in your community.

Consider the plastic sponges that are used at home. Not only does the product have the potential of leaving a lasting mark in terms of the time the plastic will remain in the environment after it is no longer needed, but the chemicals that the sponge carries also has an effect through its life cycle.

Sponges are often filled with bacteria-killing triclosan, which can negatively affect aquatic environments.[23] In addition, the 'synthetic materials that they're composed of . . . release dioxins and formaldehyde into soil and [the] atmosphere',[24] from production to the disposal of the item. Looking at this from a simple perspective, the item has caused waste during manufacturing, while you use it in your home and after it is removed from your premises. The potential of waste being created throughout a product's life cycle is one key reason why life cycles need to be evaluated in order to reduce waste worldwide.[25, 26] Instead of plastic scourers, natural alternatives such as loofahs made from sisal can be used and then composted.

An awareness of what sustainable products can be used within the home, whether they are natural repelling agents that clean your home[27] or self-cleaning paint,[28] will assist in living a more sustainable life. Similarly, understanding solutions throughout your household is beneficial to the environment. With regards to your laundry, for example, simple changes can be made to washing loads.[29] A fuller load means there is less excess water in the machine, reducing water waste. This will also limit micro-plastic particles[30] from your clothes ending up in the waterways.[31] Another useful piece of knowledge is that unlike other products that have criteria restricting microbeads,[32] common laundry detergents are not subject to this.[33] Allowing people access to this type of information will enable you to become an active consumer who can demand change from a well-informed position.

Sustainable Practices at Home

'We are starting to witness the penalty for unsustainable lifestyles and patterns of production and consumption. As the human population is exploding, resources are shrinking. Concerns loom everywhere, from declining pollinators affecting food security to air and water pollution affecting the quality of life, and land shortage and degradation affecting both agriculture and biodiversity.'[34]

Homes are typically designed without sustainability in mind. Instead, most houses, apartments, etc., are built in the most economical way possible[35] without looking at the overall life cycle. The stakeholders involved at this point are rarely engaged at other stages, highlighting a disconnect within the system. These stakeholders rarely perform an evaluation of the associated costs of running and operating a home. This possibly means that a home being built does not utilize modern resources such as solar power or green roofs, or take into consideration building designs[36] that are more sustainable in the long term because economical gains in the short term are valued higher.[37] Re-evaluating this process will enable homes to operate within a circular economy. That way the entire life cycle can be taken into account.

Despite these practices, there are numerous sustainable options that can help limit waste at home. These include energy-efficient appliances and light bulbs,[38] using local and/ or recycled materials and managing water usage.[39] All of these areas become more valuable if they are looked at as part of a system that is aware of the life cycle of other items that function side by side with one another.[40] To fully understand the waste that is produced in your home, being aware of why certain items function the way that they do, which is by looking at its life cycle, will assist you in making more of a difference to the levels of waste that you produce.

What Can You See in Your Waste Now?

'I imagine that you're getting the hang of this by now. We're progressing well together and have come a long way since we first started in the bathroom. Well done!

Each step is linked to the next and we've been looking at areas close to home so far. The next few topics will be a little broader. I'm looking forward to it, and you should as well. I'll leave you for a moment to review your lessons so far and meet you in the next chapter with a gift or two, from me to you.'

LOOK A LITTLE DEEPER WITH THIS ACTIVITY!

Activity #3 is a four-part process, on separate question sheets, that builds on everything you have learnt in this chapter. The importance of this exercise is to understand and become aware of the processes that can help you transition to a sustainable lifestyle.

ASSESS YOUR
HOME CARE WASTE:

What type of environmental impact do your products/services have?
Join the product/service to the problem using an arrow:

Product/Service: **Problem:**

Detergents   Lack of affordable
                                                                  sustainable options

Electricity   Water
                                                                  contamination

Construction
materials   Fossil fuel derived
                                                                  energy use

Use the ideas from this sample to assess your waste. The product/service
may relate to more than one problem. Fill in the lines below with your
products/services and environmental problems.

_____   _____

_____   _____

_____   _____

_____   _____

_____   _____

When thinking about the environmental impact, it is important to think of the number of areas that it could
affect. Start on a small scale and work your way out. First, think about what it means to the micro-
environment around you, then think larger and larger until you look at it from a global perspective. It will be
beneficial to undertake some research online or in a library.

ASSESS YOUR
HOME CARE WASTE:
ACTIVITY #3 QUESTION 2 OF 4

What system issues prevent change?
Join the product/service to the issue using an arrow:

Product/Service: **Issue:**

Detergents   Landfill waste

Electricity   No environmental
                                               policy in place

Construction
materials   No policy to
                                               decarbonize

Use the ideas from this sample to assess your waste. The
product/service may relate to more than one issue. Fill in the lines below
with your products/services and environmental issues.

_____   _____

_____   _____

_____   _____

_____   _____

_____   _____

A system is anything associated with the product or service that is interconnected with it, for example, the
manufacturing unit that produces the product or the government office that provides the service. To learn
more, conduct research online or in a library.

https://www.bing.com/search?q=Bare+Necessities+Zero+Waste+Solutions+Pvt+Ltd

ASSESS YOUR
HOME CARE WASTE:

What sustainable options are there to replace it?
Join the product/service to the solution using an arrow:

Product/Service:

Detergents

Electricity

Construction
materials

Solution:

Reclaimed building
material

Home-made
alternatives

Renewable
electricity

Use the ideas from this sample to assess your waste. The
product/service may relate to more than one solution. Fill in the lines
below with your products/services and solutions.

_____   _____

_____   _____

_____   _____

_____   _____

_____   _____

A sustainable option is a product or service that will last longer and/or produce less waste. Think about
options such as products made from earth-friendly materials, or those that reduce waste through a supply
chain. To learn more, research online or in a library.

How will you start using the sustainable option?
Join the product/service to the action using an arrow:

Product/Service:

Detergents

Electricity

Construction
materials

Action:

Use reclaimed
wood for homes

Make your own or
buy chemical-free
options

Implement solar
panels

Use the ideas from this sample to assess your waste. The
product/service may relate to more than one action. Fill in the lines below
with your products/services and actions.

_____   _____

_____   _____

_____   _____

_____   _____

_____   _____

This last step is all up to you. Make your choice, know the benefits and live a zero-waste lifestyle. You are more
likely to succeed with support from your friends and family.

ZERO—WASTE LIBRARY

So-da-mn versatile

Here's how to make your own floor cleaner.

You will need:

- ¼ cup of white vinegar,
- ¼ cup baking soda,
- 1 teaspoon liquid dishwashing soap or bio-enzyme, or reetha concentrate (optional),
- 2 mugs of warm water,
- A few drops essential oils (tea tree or orange).

For this all-purpose, natural floor cleaner, what you need to do is:

- Add the baking soda, vinegar and soap/bio-enzyme/reetha concentrate into a spray bottle,
- Pour in the warm water,
- Add in a few drops of essential oil,
- Secure the bottle and shake well.

There! Your very own floor cleaner that can oust those stubborn stains in a jiffy is ready.

Ironman/Ironwoman

To clean your iron, you will need:

- 1 cup vinegar (diluted in water),
- A pinch of baking soda.

Gather your ingredients and follow this method:

- Soak a cloth (or rag) in vinegar,
- Sprinkle baking soda on the cloth,
- Place the iron on the cloth and move in a circular motion (with the iron switched off),
- Once it's clean, turn on the steam setting and let the baking soda fall off,
- Repeat if necessary.

This is super simple and super useful!

Out of Odour

To deodorize your musty upholstery, you will need:

- Baking soda.

This is such a simple trick to not just deodorize your musty upholstery but also your couch, pet beds, your own bed and many other things around your home. All you need to do is:

- Sprinkle baking soda on the area you want to treat,
- Allow it to sit there for 15 minutes,
- Use a vacuum cleaner.

The odour should disappear. If not, add a little more and repeat.

Clearly Cranky With Crayons

To remove crayon marks from your wall, you will need:

- Baking soda.

This one is a must if your toddler has drawn on the wall! Follow this method to clear it all up:

- Sprinkle the baking soda on to a damp cloth, rag or eco sponge.
- Scrub it all away!

This treatment should clear the mess away (best not to tell your kids the secret though, just in case they have another go at it).

Soap-er Nuts
This is a multipurpose laundry, floor and/or dishwashing liquid made using soap nuts.

You will need:

- 2 litres of water,
- Soap nuts.

The method will give you the base liquid for a home cleaner.

Start with:

- Adding 20 soap nuts to a pot of water,
- Boil the water with the soap nuts,
- Once boiled, carefully extract the soap nuts and deseed them,
- Place the sleeves back into the boiling water and boil for a few more minutes.

Once the water has changed colour, allow it to cool down. This is now the base liquid for a number of cleaners. All natural, of course!

Natural Tide

Here's another way to make a natural floor cleaner.

You will need:
- 3 cups reetha liquid,
- 2 cups white vinegar,
- 2 tablespoon pink Himalayan salt, or rock salt,
- 20 drops of citronella (great to keep the mosquitos away) or tea tree essential oil (a great disinfectant).

All you need to do is:

- Mix the reetha liquid and vinegar together,
- Add salt,
- Stir it all together.

Your floor cleaner is ready to use. Keep in mind that if you need less than the quantities this makes, you can store it in the fridge.

Sunshine Tide

This is how to make a handy dishwashing detergent.

You will need:

- 3 cups reetha water,
- 1 cup glycerine,
- 1 teaspoon (or as much as you need) Xanthan gum (a thickening agent),
- 3 drops lemon essential oil,
- 3 drops tea tree essential oil.

This one needs a blender. Gather all the materials and ensure that the all essential oils are pure and cold pressed. Follow the method to get an amazingly clean floor:

- Blend together the reetha water, glycerine, Xanthan gum and essential oils,
- Pour it into a bowl.

It's now ready to use!

< 135 >

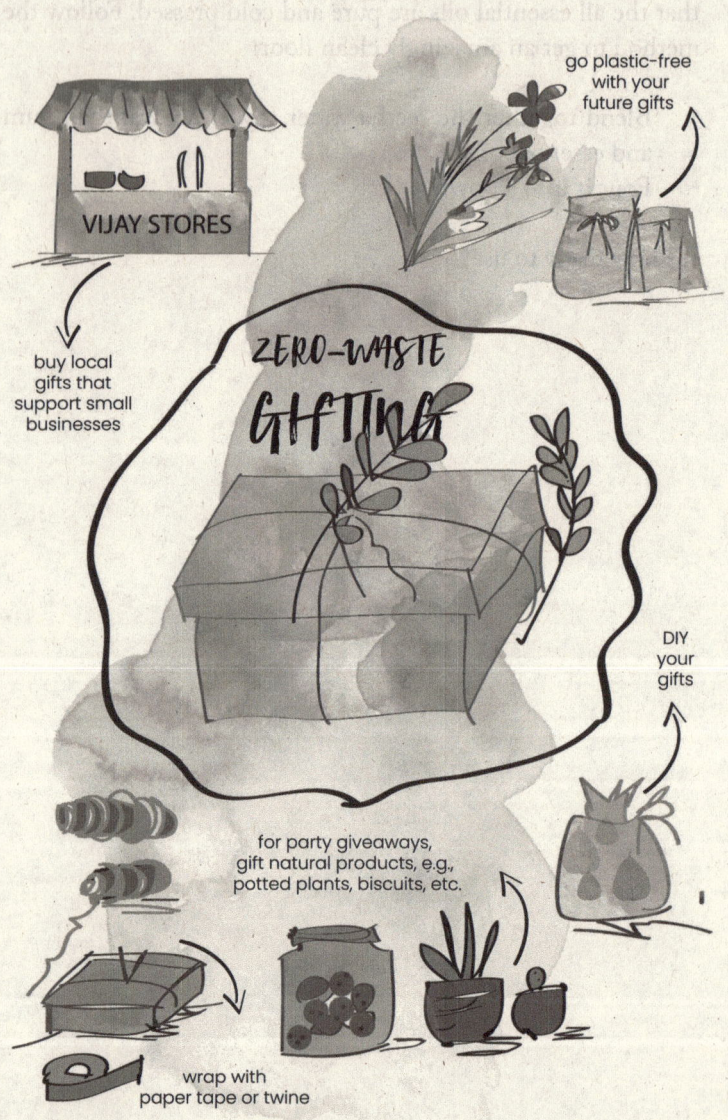

VIJAY STORES

go plastic-free with your future gifts

buy local gifts that support small businesses

ZERO-WASTE GIFTING

DIY your gifts

for party giveaways, gift natural products, e.g., potted plants, biscuits, etc.

wrap with paper tape or twine

CHAPTER 5

GIFTING

'We celebrate our festivals with great pomp and show, but we also add a lot of waste and leave no stone unturned to contribute immensely to air pollution. Given the poor quality of air that we breathe day in and day out, it may come as no surprise that the World Health Organization has categorized air pollution as the sixth biggest cause of deaths in India, triggering an alarm, with studies showing breathing ailments on the rise in Indian cities.'[1]

The Perfect Life . . .?

You love this time of the year. There are celebrations all around. The night before Diwali, fireworks are ablaze throughout the city. There is so much noise, so much warmth. It is wonderful. You have already made a plan with your friends to go back to the store to get more crackers because you ended up using more than expected on the first day of the holiday. You're more than happy to splurge and make that extra purchase.

You have gifts for all of your family members on the table at home. You wrap some of them up in a dazzling blue material, a couple of the smaller ones in glitzy purple and pink, and the remaining with the red, binding them all with the pretty plastic string ribbons that you bought. You place them all in one bag while the additional wrapping material that you brought back from the large shopping centre, with its amazing selection of gifts, is placed into the bin.

You do hope that your family enjoys the gifts. You put a lot of thought into them. Although the planning has not been fool proof with some of your gifts in the past, you feel like it will work today. There should be no complaints, you think.

Next, you move to a smaller pile. They are your gifts for your friends. You wrap them in blue, purple and pink, and one or two in the remaining wrapping of red. You have finished a number of the wraps that you chose at the shopping centre yesterday where one of the big stores, a fancy global brand that has just opened, was offering terrific deals. All of the other gifts came from similar stores. You do love the way that everything is standardized, letting you find anything you want to these days.

You are an amazing gift-giver, you think. Far better than some of your friends and family. There have been so many times that you've had to throw those useless things away. So many gifts over the years with no purpose. What a shame! You are certain that none of your gifts will be discarded. Anyway, even if that happens, it must be because of the receiver's inability to see a good and useful gift. You put a couple of hours, after all, into shopping for all of it at the shopping centre. They should appreciate that!

What Is in Your Waste?

'India's estimated ten million weddings a year contribute significantly to its annual food waste, worth about US $14 billion [Rs 1400 crore approximately] in losses.'[2]

> 'Gift-giving has become almost essential during holiday seasons all around the world. Yet, the amount of waste this practice produces is mountainous. Simple choices, such as using fireworks to celebrate or wrapping items in single-use materials, have far-reaching implications on the environment. The way gifts are given today has led to materials being discarded with an ease that was rare in yesteryear. This has led to consumers feeling detached from the gifts that they receive, which has increased the potential of them being thrown away.
>
> I have some really cool stories and zero-waste recipes to share with you about all of this in the next couple of sections. Keep in mind some of the earlier discussions we've had about food waste. Festivals

can produce truckloads of waste. Becoming aware of all of your waste is the first step to start reducing your impact.'

NOW IT'S YOUR TURN TO ASSESS!

Try out Activity #1 with your current knowledge. View this activity as an initial way to see how much you know. Once you learn more as we go forward in this chapter, you can return to it in order to see how much your knowledge has developed.

ASSESS YOUR
GIFTING WASTE:
ACTIVITY #1

Follow the example sheet below by choosing a few of your own products to assess.

	Example A	Example B	Example C
Is the container recyclable or reusable?	✓	✗	✓
Is there a risk that the product will have a large environmental impact?	✗	✓	✗
Are there chemicals/residual impacts, like smoke from fireworks, that can harm people and the planet?	✗	✗	✓
Could you gift an experience or a handmade product to reduce wastage?	✓	✓	✗

- If yes: share the solutions with friends and family.
- If no: research options.

Use the ideas from this sample to assess your waste. You can draw your own sheet based on this and create other questions for this assessment based on your needs.

What Resources Are Available?

'There has always been a sense of fun associated with gifting for me. There is also a lot of nostalgia when I think about many of the gifts that I have received over the years. The ones that I hold the most attachment to have been made by hand or remind me of cool experiences that fill me with joy. There was this one time, for instance, when I received a tree!

I find the entire concept of planting a tree grounding. It is a gift that keeps on giving back for a long time. Fruiting trees, in particular, have shown me the value of this. When I was younger, my dad and I planted a coconut tree as a gift in my family backyard. The crazy thing is that it grew to a full-sized glorious tree but only started fruiting a decade after he passed. It provides us with thirst-quenching coconut water to this day. This shows that gifts in the form of experiences, like planting trees, continue to nourish us with happy memories for a long time, over and over again, helping us form an even deeper connection with the person who gave us the gift, in this case, my dad.

I will be forever grateful for that sunny Sunday afternoon when this seed was planted. The value of such a gift has always been more important to me than receiving something like a pair of jeans or a shirt. I'm sure you have some great stories like this, too. I hope I get to hear about them some day.'

TRY THIS ACTIVITY OUT!

In this chapter, there are a number of suggestions that you can use for Activity #2. You can read on and return with new knowledge, or assess yourself now before learning a little more and then return to this exercise.

ASSESS YOUR
GIFTING RESOURCES:
ACTIVITY #2

Create a zero-waste gift for a friend or family member and/or gift an experience. Use the suggestions in the guide book or research to find more. Fill in each section of this activity to record your achievements.

The resources that I have learnt about are:

 'A Peanut Butter Surprise To Spread Some Love' recipe

 'Don't Leaf Me Up' activity

 I took my best friend for a nature walk, through a forest and all the way to the beach.

Draw a picture from your activity to show friends and family:

The resources needed for this activity are:

 Peanuts, peanut oil, salt, dry fruits and a cool-looking jar to gift it to my sister.

 Leaves, earth-friendly paint and string to hang the decorations for a dinner party.

 We took public transport to the beach, a reusable bottle for water and enjoyed our time.

Record where you learnt about these resources:

 The guide book's 'Zero-Waste Recipes' section, 'Zero-Waste Library' and online research.

Record who you shared your success with:

 My sister who loved her surprise! My best friend! We had so much fun on our walk.

Use the above ideas to assess your resources and then create your own activity sheet.

Moving Towards a More Sustainable Lifestyle

'Thanks to the memories I have from when I was younger, I was able to fall back on those learnings to ensure that all the gifts I give are sustainable. Looking at the resources around me that could be reused, like old clothes that I could sew into a quilt, and a bit of online research for sustainable ideas, made it a very achievable process. It meant that I could transition to sharing zero-waste gifts simply by being conscious of my own desire to do so (I've detailed a few fun recipes below and included more at the end of the chapter. Gifts from me to you!).

Experiences, too, hold a lot of value as an alternative that won't add to waste and be memorable for a long time to come. One of my favourite gifts was from my best friend after I graduated from college. It was scuba-diving lessons! I'll never forget that. However, it doesn't need to be something expensive, like those lessons were, to be sustainable. It could be a walk in a park with a loved one or simple DIY gifts like the three listed below: home-made peanut butter, a cookie mix jar and a soy candle.'

ZERO—WASTE RECIPES

A Peanut Butter Surprise to Spread Some Love
Here's your own DIY peanut butter recipe.

You will need:

- 400 gram of roasted, lightly salted peanuts,
- Peanut oil or any neutral-flavoured locally sourced oil,
- Salt,
- Assorted dry fruits for garnish.

This recipe is really easy and really yum! All you need to do is:

- Add peanuts to a blender,
- Add 1 tablespoon of oil,
- Add a pinch of salt,
- Blend for 1–2 minutes until it has a nice, smooth consistency,
- Spoon the mix into a clear jar,
- Add an assortment of dried fruits,
- Seal the lid.

For gift-giving purposes:

- Place your choice of sustainable decorations on the outside of the jar, or on the lid, like a string with a cardboard note saying 'Happy Birthday!'

All The Oatmeal Chocolate-Chip Cookies You Can Munch On!
This is a DIY cookie mix.

You will need:

- ½ a cup of quick or old-fashioned oats,
- ⅓ cup brown sugar,
- ⅔ cup multipurpose flour,
- ¼ teaspoon salt,
- ⅓ cup jaggery,
- ¾ cup chopped nuts,
- ¾ cup dark chocolate.

The way I like to give this gift is to place all the ingredients in a clear glass jar, instead of stirring it up in a bowl, and baking it myself to take to my friend's place. All of the layers really have a cool visual effect. It is a big wow factor!

The rest of the recipe is designed for those who chose the wow factor! (If you haven't, simply skip the step involving the jar and head to the baking stage). Here are the items you need to bake:

- ½ cup butter,
- 1 large egg,
- ½ teaspoon vanilla extract.

Gather everything (the cookie mix you brought and the extra items to bake). Make sure to preheat the oven to 190°C. Then:

- Beat the softened butter, egg and vanilla extract in a large mixer bowl until it has blended well,
- Add cookie mix,
- Mix well, breaking up any lumps,

- Drop the mixture using a rounded tablespoon on to ungreased baking sheets (you should have 2 dozen cookies),
- Bake in preheated 190°C (375°F) oven for 8–10 minutes,
- Remove and cool on baking sheets for 2 minutes.

For the gifting stage, remember:

- Fill the jar with your cookie mix,
- Place your choice of sustainable decorations on the outside of the jar, or on the lid, like a string with a cardboard note saying 'Congratulations on Your Book Launch!'

'All of these creative elements remind me of my childhood with my sisters, with my mum encouraging us to be active and imaginative. I loved those days. I'm sure you have some amazing memories of your childhood experiments, too! I really think that, like a lot of areas in our lives, to be sustainable we need to look back at the simple, creative and fun methods that we used. I don't want to sound repetitive by saying this, but think about cooking a meal as a gift, making handmade cards, planning a picnic in the park and other thoughtful gestures. It is a sense of expression and appreciation for our loved ones, wrapped in sustainable ideas.

I think we can all learn lessons from baking a cake, where you need to be precise with the ingredients and aware of all the processes. It is all about slowing down, taking time and being more mindful. We can do that for all our festivals and gifting processes by looking at how they have changed and understanding the reasons why this has happened.

Take, for example, the firecrackers on Diwali that cause immense air pollution. This is a modern problem that causes excess waste due to how we choose to celebrate the festival. If all of us reflect a little when we have time to spend with our loved ones, we will find better methods of celebration so that no one is harmed, be it us or the planet. One great move

< 148 >

is to light diyas (pro tip: use excess cooking oil) instead of bursting crackers, like we traditionally did. Growing up, I saw biryani being exchanged on Eid and getting Christmas goodies and sweets for Diwali, which really encouraged sharing between neighbours from different cultures.'

A Candle Light From My Heart to Yours . . . Smelling Like Citronella!

This is how you can make a DIY vegan candle.

You will need:

- A glass jar,
- Soy wax,
- A wick,
- Scissors,
- Natural fragrance oils of your choice (my recipe has citronella),
- Two pencils.

Follow this method, but remember to be careful with the hot wax:

- Measure how much wax you need to fill the jar to an appropriate level,
- Pour the wax into a double boiler,
- Allow it to melt for 10–15 minutes,
- Add the scented oils,

- Cut the wick to an appropriate length so that a good amount is above the wax when it is in the jar,
- Dip it into the melted wax,
- Wait until the wax hardens.

The finishing touches will take a little time. All you need to do is:

- Hold the stiffened wick inside the glass jar,
- Slowly pour the wax into the container,
- Rest two pencils on the top of the jar to hold the wick in place until the wax hardens (you can do this with your hands, too, but it may get tiring. I find having a wick held in place between the pencils is less effort!),
- Allow the wax to set for 4 hours at room temperature.

For gift-giving purposes:

- Place your choice of sustainable decorations next to the outside of the jar or on to the lid like a string with a cardboard note saying 'Happy Diwali!'

'Other amazing ideas for sustainable celebrations include wooden Christmas decorations such as Channapatna toys, which can hang on a Christmas tree in the form of cool and beautiful ornaments, or on a windowsill next to other decorations. Similarly, during Eid, my sisters and I used to make greeting cards for our elders and get "Eidi", which is money from elders, in return. My amazing mama opened a bank account for me and deposited all the Eidi I received over the years in it. When I was eighteen, I had enough to buy myself a laptop for college!

Making home-made gifts allows you to slow down and be mindful of everything you create. Whether it is a spaceship-shaped cake, or a card that your mum will cherish forever, sharing and appreciation has always been the heart of our culture. These days, the way we do it has changed dramatically.

I think, with gifting becoming such a norm, it is easy to forget the true values that they stand for. Acknowledging and embracing the thought behind why one gives gifts is far more important if we are to value gifts and relationships at a deeper level. If we do this, we can minimize the number of items that end up in landfills, such as mass-made consumables bought at major retailers.

Changes like these can make a tangible difference if we think back to how we once celebrated, while becoming aware of the impact our gifting and festive practices can have on us and the environment, be it in the form of air or water pollution.

With large, conveniently located retail outlets and online ordering, accessibility and ease of purchasing a gift is definitely not an issue any more, especially when both parents have a job. A lot of times, this ease can lead to the sentimental value behind a gift being lost. I think it is a great move forward towards a more sustainable lifestyle to use your fond and creative memories as an inspiration and share it with your children. On the other hand, if you choose to buy a gift from a shop, there are sustainable and ethical businesses that have on offer really creative gifts. All we need to do is put some thought behind giving gifts and using the best parts of growing up and being imaginative with the realities of how busy we are today.

Zero-waste gifting is a chance to be mindful, loving, caring, creative, nostalgic, memorable, imaginative, sustainable, fun and active . . . Need I say more?'

ZERO—WASTE TIPS AND TRICKS

- Buy local gifts that support small businesses and enterprises,
- Go plastic-free with your future gifts, such as flowers that are not wrapped in cellophane, (or better still, gift seeds or a plant!),
- Use cloth bags or fabric to wrap gifts instead of plastic sheets,
- DIY your gifts (crocheted mittens, a scarf, or blanket, or make some delicious jams, cakes or candles),
- Wrap with paper tape or twine instead of plastic tape,
- Gift experiences instead of material items,
- For party giveaways, gift natural products, e.g., potted plants,
- Have friends and family donate money to your favourite charity as a gift,
- You could give tickets to something fun (dance lessons, pottery lessons or concert tickets, perhaps),
- Gift a zero-waste starter kit (straws, beeswax food wrap, comfortable bamboo toothbrushes, seed pens and pencils),
- Travel e-vouchers through online travel sites,
- Gift membership or subscription to something they love, perhaps a library, audio books, or a podcast or music-streaming website or app.

GIFTING - DISTRIBUTION TO CONSUMPTION

waste produced
from large retailer

impulsive buying encouraged to
sell more for immediate
profit for large chains

waste at home
ends up in
landfills

pollution through
transport

single-use
products valued

large festivals can add to
pollution through practices
such as bursting crackers

GIFTING – DISTRIBUTION TO USE AS A RESOURCE

run on
renewable energy

consumers support
artisan livelihoods

materials can be
reutilized
(no waste)

artisans providing products
that benefit their
livelihood using
earth-friendly materials

multi-use products
or those made from materials
that can be reutilized

gifting experiences
and utilizing
sustainable transport

TICKET

©Bare Necessities Zero Waste Solutions Pvt. Ltd.

Why Is It Harmful?

'So far we've learnt about the many positives associated with gifting sustainably, but like the chapters before, it is really important to understand why our choices can have positive or negative impacts on the world around us. We'll be looking at single-use products and packaging in this section, as well as the pollution (another word for waste!) witnessed during festive periods. By learning about this, my hope is that you'll know how to make environmentally friendly decisions.'

Whether it is Diwali in India, Christmas in the USA, Eid in Turkey or New Year's Eve in Japan, celebrations and gift-giving are often an integral part of the festivities. Yet, with this comes the potential of waste. This can be in the form of gifts that are not used for a long time after being received, not really knowing what the person you're gifting really needs, single-use wrapping material or boxes, food that is uneaten because too much is prepared[3] and smoke from firecrackers, to name a few.

Single-Use Products and Packaging

'Most wrapping paper for presents can't be recycled. It's bound for the landfill.'[4]

In the USA, between Thanksgiving and New Year's, it is estimated that the amount of waste increases by 25 per cent, or 1 million extra tonnes per week. Annually, estimations on gifting discards range from over 61,000 km of ribbon to

$11 billion USD on packaging.[5] Similarly, in the UK during Christmas, an average of 83 km worth of wrapping paper ends up in dustbins.[6] Other extreme amounts of waste are produced during the Chinese New Year and Diwali.

Perhaps one of the most profound examples of waste is in India, 'a country obsessed with grand celebrations'[7] and has an estimated '10 million weddings [taking] place in India every year and all of these celebrations leave behind truckloads of trash in the form of discarded plastic cutlery, used flowers and a large amount of food . . . On an average, an Indian wedding hosting 400–1000 people results in around 3 tonnes of waste, and in some cases even more.'[8]

The underlying factor behind the increase of waste because of gifting and festivals is an increase in consumption,[9] irresponsible/naive consumerism and/or a personal view wherein one believes that such occasions need to be elaborate. Upsurges in purchases during festive periods, where gifts are exchanged en masse, provides a profound case in this situation. The underlying fact is that waste is produced due to social desires and norms.[10] Establishing what is likely to become waste and what is likely to be used as a resource after gifting is of paramount importance.[11] For instance, a thin film of wrapping plastic is unlikely to be used again, while a glass container may be.

Recyclable products are useful, but they have limitations, too. For example, stopping a product or packaging from ending up in landfills is reliant on the consumer assisting in the recycling process, through segregating and providing the product to a recycling centre.[12] Further, the recycling centre needs to have the capability to recycle the type of material provided. Clamshell plastic[13] and blister packs[14] are prime examples of recyclable materials that are likely to end up in landfills. This is particularly because the material, despite being made of PET (polyethylene terephthalate) plastic,[15] is often not collected for recycling and is not used by the consumer once he or she obtains the product inside.[16] Difficulty in recycling packaging leads to thin plastics and other less valuable forms of plastic[17] being less likely to be recycled than PET.[18]

As with many of the areas in this guidebook, recycling will also require broader change in society. Thinking about the life cycle of a product before, during and after a consumer uses it is required for items to become resources. Overall, this will often mean an increase in the transparency of business supply chains, higher levels of conscious consumerism and a shift away from the current, widely used materials. There is an inherent economic issue with the current process due to the linear economy, where it makes more economic sense to use virgin materials (a short-term outlook) instead of investing in recycling processes that may benefit the environment (a long-term outlook) as well as continue to benefit the economy and people.

From production to consumption to disposal, the potential of gift wraps, packaging and/or products turning into waste is extremely likely. This is because of the disposable process that mass linear consumerism promotes.

In order to limit the risk of the receiver undervaluing a gift, research has found that the more people are aware of the type of gift a person would like to receive, such as an experience or a need, leads to better choices overall. 'The reason experiential gifts are more socially connecting is that they tend to be more emotionally evocative . . . An experiential gift elicits a strong emotional response when a recipient consumes it— like the fear and awe of a safari adventure, the excitement of a rock concert or the calmness of a spa—and is more intensely emotional than a material possession.'[19] Poignantly illustrated is the current transition to sustainability that consumers are moving towards worldwide.[20]

Across the world, there are now events that aim to limit waste, changing how festivals and/or days that have traditionally incorporated gifting, are attended. In India, the city of Raipur in Chhattisgarh hosted a four-day 'Garbage Festival'.[21] In Bengaluru, Karnataka, Echoes of the Earth,

'India's first and only "green" (music) festival',[22] has been hosted multiple times. Weddings, too, are moving towards a zero-waste direction[23] in the south Asian nation.[24, 25, 26] Countries such as Italy, USA, Australia, Slovakia, Iceland, The Netherlands, Denmark, France,[27] and the United Kingdom and Portugal[28] are seeing a positive shift in mindsets, the willingness to talk about a previously unrecognized (or simply ignored) topic and an awareness of the larger impact our choices during festive occasions can have.

Pollution Created During Festive Periods

*Fireworks cause extensive air pollution in a short amount of time, leaving metal particles, dangerous toxins, harmful chemicals and smoke in the air for hours and days. Some of the toxins never fully decompose or disintegrate, but rather hang around in the environment, poisoning all they come in contact with. Exposure to fine particles, like those found in smoke and haze, is linked to negative health implications, such as coughing, wheezing, shortness of breath, asthma attacks and even heart attacks.'[29]

In addition to gifts, pollution during festive periods is commonplace, e.g., because of lighting fireworks.[30] An increase in air pollution because of the smoke from these crackers has been associated with cardiovascular and respiratory mortality and morbidity.[31, 32] The maximum extent of this type of pollution was recorded after Diwali in 2019, in the Indian capital New Delhi where the Central Pollution Control Board's measurement of the air quality index[33] showed the highest ever readings.[34] This was said to be a profoundly dangerous situation for all residents, particularly the underserved communities who live in informal housing.

While the Indian capital gets all the media attention for the negative impacts of air pollution, rural areas, such as one studied in northern Haryana, have also recorded 'soaring levels of PM2.5 and carbon monoxide'.[35] So do other locations, including Chennai, Kolkata and Mumbai, and smaller cities of Bhiwadi in Rajasthan, Agra in Uttar Pradesh and Ludhiana in Punjab.[36] However, it is not only in India that the impacts of bursting firecrackers is felt. 'China, the home of fireworks, also sees sharp spikes in pollution after big celebrations,[37] with neighbouring country South Korea recording poorer quality of air after 'distant lunar New Year's celebrations in China'.[38]

There is no blanket rule to say that one gift, or festive celebration, is better than another. However, what is becoming clearer with each study and report is that the impacts of wasteful practices are felt far and wide. The amount of waste is growing with populations and income levels increasing in most countries,[39] which means more disposable income being available to spend on single-use celebratory items.

A tipping point has been reached. No longer can celebrations claim that the waste produced is centralized. Whether it is in the air, on land or in waterways, waste is clearly visible. 'India's rivers have a very particular problem with pollution. As part of religious celebrations, tonnes of flowers [are] left at temples every day . . . as populations have grown and so the number of temples has increased, an estimated 8 million tonnes from somewhere in the region of 600,000 places of worship have now [been] dumped in this way every year.'[40] Additionally, traditional methods of celebration that promote the distribution of celebratory items into river systems in the populous country have led to 'toxic arsenic, lead, and cadmium . . . contributing to turning these bodies of water into a potentially deadly carcinogenic soup.'[41]

Small-scale choices can have broader effects. The knowledge you can get, from stories and information about the impacts of pollution, will enable you to view the complexity of systemic issues and address them the best you can with certain societal factors,[42] such as waste management operations, determining to some extent how much your actions make a difference.

Becoming a catalyst for change can see you start off from providing a sustainable gift, such as an experience. A broader change will occur when enough people shift their gifting habits, ensuring a linear economy is no longer in place because entire communities will seek gifts and activities that leave no indelible mark on the land, water or air. Gifting will change with a circular system, but that does not mean it won't be fun.[43] Instead, it will last for far longer, helping the majority of people transition into being conscious citizens who care for their families, neighbours and the world.

What Can You See in Your Waste Now?

'I love gifting, as I'm sure you can tell! Read through a few more of the recipes and ideas after this last activity in the chapter. They're a lot of fun. Our next stage together will take us outside of our homes. Collect your lessons, pack your zero-waste bags and join me there!'

LOOK A LITTLE DEEPER
WITH THIS ACTIVITY!

Activity #3 is a four-part process, on separate question sheets, that builds on everything you have learnt in this chapter. The importance of this exercise is to understand and become aware of the processes that can help you transition to a sustainable lifestyle.

ASSESS YOUR
GIFTING WASTE:

What type of environmental impact do your products/services have?
Join the product/service to the problem using an arrow:

Product/Service:

Wrapping paper

Party decorations
and servingware

Fireworks

Problem:

Air pollution

Plastic pollution

Landfill waste

Use the ideas from this sample to assess your waste. The product/service
may relate to more than one problem. Fill in the lines below with your
products/services and environmental problems.

_____   _____

_____   _____

_____   _____

_____   _____

_____   _____

When thinking about the environmental impact, it is important to think of the number of areas that it could
affect. Start on a small scale and work your way out. First, think about what it means to the micro-
environment around you, then think larger and larger until you look at it from a global perspective. It will be
beneficial to undertake some research online or in a library.

ASSESS YOUR
GIFTING WASTE:

What system issues prevent change?
Join the product/service to the issue using an arrow:

Product/Service:		Issue:
Wrapping paper		Respiratory illness
Party decorations and servingware		Ocean pollution
Fireworks		Perpetuating a wasteful system

Use the ideas from this sample to assess your waste. The product/service may relate to more than one issue. Fill in the lines below with your products/services and environmental issues.

_____   _____

_____   _____

_____   _____

_____   _____

_____

A system is anything associated with the product or service that is interconnected with it, for example, the manufacturing unit that produces the product or the government office that provides the service. To learn more, conduct research online or in a library.

ASSESS YOUR
GIFTING WASTE:

What sustainable options are there to replace it?
Join the product/service to the solution using an arrow:

Product/Service:

Wrapping paper

Party decorations and servingware

Fireworks

Solution:

Light shows using drones or diyas

Use recyclable or reusable material

Rent party supplies

Use the ideas from this sample to assess your waste. The product/service may relate to more than one solution. Fill in the lines below with your products/services and solutions.

_____   _____

_____   _____

_____   _____

_____   _____

_____   _____

A sustainable option is a product or service that will last longer and/or produce less waste. Think about options such as products made from earth-friendly materials, or those that reduce waste through a supply chain. To learn more, research online or in a library.

ASSESS YOUR
GIFTING WASTE:

How will you start using the sustainable option?
Join the product/service to the action using an arrow:

Product/Service:

Wrapping paper

Party decorations
and servingware

Fireworks

Action:

Celebrate with
diyas

Gift wrap in
newspaper

Make your own
decorations

Use the ideas from this sample to assess your waste. The
product/service may relate to more than one action. Fill in the lines below
with your products/services and actions.

_____   _____

_____   _____

_____   _____

_____   _____

_____   _____

This last step is all up to you. Make your choice, know the benefits and live a zero-waste lifestyle. You are more
likely to succeed with support from your friends and family.

ZERO—WASTE LIBRARY

Glued to Nature[44]

Here's how you can make your own DIY glue.

You will need:

- ½ cup water,
- 1 tablespoon white flour,
- 1 tablespoon cornstarch,
- 1 tablespoon white vinegar.

This is truly the sticky side of nature! All you need to do now is:

- Bring the water to boil in a small saucepan,
- Add the white flour,
- Add the cornstarch,
- Slowly mix it all together on a low flame and stir constantly until it becomes a thick paste,
- Take it off the stove,
- Dilute the white vinegar,
- Store in a small glass jar and apply with a paintbrush (with natural bristles, preferably).

It dries clear and works perfectly for art and craft purposes.

Rollin' in the Art

To make your own DIY pen stand, you will need:

- An empty toilet paper roll,
- Paint/colour/doodling stationery (earth-friendly, of course),

< 165 >

- Kraft paper,
- Glue (try out the glue recipe in this book to make some yourself!).

This gift will bring out your creative side. Follow this method (or let your creativity run free):

- Hold the empty toilet paper roll,
- Use your imagination to paint/colour/draw on it,
- Or take blue or white paper, and cut it to fit the roll,
- Paste it on the roll,
- Draw a little unicorn, cat, a little boy or girl, or maybe even a few snowflakes on it,
- Take a small piece of paper,
- Place the toilet roll over it,
- Take a pencil and trace the circumference of the roll,
- Draw a little square beyond the boundary of the roll,
- You can now erase the circle or simply turn it over,
- Use glue to stick the ends of the toilet roll on to the paper,
- Let it dry for 2 hours.

When you've finished all of those steps, you can place your pencils into it. There you have it, your perfect pencil holder!

Tin That Can into Art

This is an easy way to make a DIY planter or pen stand.

You will need:

- An empty tin,
- Paint (earth-friendly, of course),
- Old gift-wrapping paper,
- Glue (you can make some yourself using the glue recipe).

< 166 >

Now that you've gathered all the items, the next step is to:

- Place the tin on a table or a bench,
- Use your imagination to paint different shapes, styles and patterns on to it, or wrap it up with the gift paper instead,
- Paste it over the sides of the tin,
- Draw more patterns on the paper wherever you like.

Place your pencils inside (or you could use the same method to decorate a planter).

Don't Leaf Me Yet!
Now you can make DIY leaf decorations for the next party you host.

You will need:

- Dried leaves,
- Paint (earth-friendly, of course),
- String.

This one is for your natural, artsy side! Here's how to do it:

- Collect some dried leaves from your garden,
- Paint a bunch of pretty patterns on them,
- String them together to enhance the aesthetic of your room, or hang them elsewhere, like on a box or your cupboards.
- You could also use them to cover gifts, or as decorations at a dinner party!

When you're all done with the leaves, they can be placed into your composter.

Make-Your-Own Bamboo Speakers[45]

You will need:

*   Bamboo, approximately the size of your two hands together (roughly 10 cm), however, you can make it to suit the size you need. Ensure that it is compatible with the size of your smartphone or other device.

Follow this method:

*   Cut the bamboo to the desired length, but ensure that it is still long enough to shape, roughly 5 cm.
*   Clean the inside of the bamboo using a file and/or sand paper,
*   Cut both sides of the bamboo, at an angle of approximately 30–60 degrees,
*   Next, sand the bottom of the speaker,
*   Drill a pair of holes 2 cm at the top. The slot should be long enough for your smartphone,
*   Use a sabre saw to cut a hole that will fit your smartphone,
*   Next, drill 2–3 holes, 2 cm big, in the front. This will help project the sound better,
*   Sand the entire surface to get it ready,
*   You can also paint your speakers or add other patterns in other earth-friendly ways.

Zero-Waste Party Rentals[46, 47]

Here is a list of individuals and organizations promoting the zero-waste movement in India. These party rentals are truly useful, inspiring and amazing!

*   Namma Cutlery Bank by Namma Ooru Foundation, located in Bengaluru,

- GreenishORA, located in Pune,
- Rent-A-Cutlery, located in the Sarjapur and Whitefield areas of Bengaluru,
- The Sophisticated Touch, located in Panaji, Goa,
- ReUse Cutlery Bank, located in Bengaluru,
- Hamdubhai Mominbhai & Sons (HMS), located in Ahmedabad,
- Adamya Chetana, located in Bengaluru, Hubballi, Kalaburagi and Gadag in Karnataka,
- SpillSavers, located in Bengaluru,
- Aswath Narayana (MLA) Steel Bank, located in Bengaluru,
- Hamsa, located in Bengaluru,
- Crockery Bank for Everyone, located in Gurgaon,
- Trash Trimmers, located in Indirapuram and Ghaziabad in Uttar Pradesh/NCR. Also located in Delhi,
- The Bartan Company, located in Hyderabad,
- Seeth Anand, located in Hyderabad, Flimnagar, Jubliee Hills, Gachibowli and Tolichowki in Telangana,
- Reena, located in Aparna Sarovar, Hyderabad,
- Sushoma, located in Aparna Cyber Zon, Hyderabad.

CHAPTER 6

COMMUNITY

'In Australia, the e-waste problem is compounded by the fact [that] we don't consistently direct our used devices to formal recycling centres. India has that plus the added problem of 'backyard' recyclers who mine the streets for discarded e-waste and use dangerous methods to extract the elements within, causing potential harm to themselves and the environment.'[1]

The Perfect Life . . .?

The traffic was bad this morning, but you made it on time. You enjoy being here. The fact that you have to deal with traffic most days of the week is of little consequence because you have a new radio system in your car. Now that you have reached you can work on your project; it is a big one this time. One that will take multiple hours and involve many colleagues. Fortunately, there has been a recent upgrade with phones so that anyone can reach you at any time. You had a good model earlier, but it was a year old and needed an upgrade. It was around the same age as your laptop, which reminds you that there is a sale on laptops at the nearby computer shop on the weekend.

You make a stop at the cafeteria and buy a coffee and a small muffin to-go. You drink the coffee on your way back, spilling a little on your shirt, and gobble the little banana muffin. You have walked a fair amount and are too far away from the cafeteria. It's too late to go back to the segregated bins so you discard both, the cup as well as the muffin wrapper, into the same rubbish pile that some people have started by the wall near the elevator.

Entering your room, you sit down on a seat you prefer. It is not the same as the large individual room that you used to use, and which you prefer, but this works too. You have already placed a request to get a room for yourself again. That seems like a better utilization of space; these multi-capacity rooms are a distraction. You pull out your pen and paper, only to find that the pen is no longer working. You pull another

< 172 >

one out from your bag and that, too, has the same issue. You push both of them to the far side of the table for the cleaner to remove later.

You switch on your computer and it whirs back to life. You go about your day. You call people on your new phone and email your colleagues. At lunch, you order from a delivery company and eat while you work. The pile for the cleaner becomes a little larger. You think about talking to someone regarding that as you add an empty water bottle and your second takeaway coffee cup to the pile.

By late afternoon you have made a lot of progress and feel that you should buy that new laptop for yourself, as a well-deserved reward. You leave the room. The pile of waste still sits there under the glow of the lights and the luminescence of other computer screens that were left on. You shake your head at the cleaner's inefficiency and walk to your car. For a moment, you wonder if you will have a traffic-filled drive home, but then a great song plays on the radio just as you turn on the engine, making you forget all of your concerns.

What Is in Your Waste?

'Waste generation rate depends on factors such as population density, economic status, level of commercial activity, culture and city/region.'[2]

'The sense of community that we have in our country is something we need to celebrate. I really feel that as a country grows and develops, people have a tendency to forget about the community. To me, community does not only mean the people that you live with or the small block in your city or town that you spend time around. Instead, I see the community as different aspects of society.

It can include the people we work or study with and it can include the people we commute with each day.

This interconnected nature of communities forms a very beneficial relationship with circular economies.

Examples of the unique Indian sense of communities are bath houses, community centres, courtyards, community streets where kids play cricket, chai shops and bazaars. Knowing how and why waste is generated in these areas (you can use the lessons from the earlier chapters to work this out) will allow you to work out what type of waste you can help reduce in each case.'

NOW IT'S YOUR TURN TO ASSESS!

Try out Activity #1 with your current knowledge. View this activity as an initial way to see how much you know. As you learn more as we go forward in this chapter, you can return to it in order to see how much your knowledge has developed.

< 174 >

ASSESS YOUR
COMMUNITY WASTE:
ACTIVITY #1

Follow the example sheet below by choosing a few products/services (for services it could be something like transport) you use to assess.

	Example A	Example B	Example C
Is this the most sustainable transport option available?			
Is the product/service harmful to the environment?			
Does the product promote unsustainable practices, such as a need to upgrade phones/laptops regularly?			
Do you know the supply chain of a product/service that you are using?			

- If yes: share the solutions with friends and family.
- If no: research options.

Use the ideas from this sample to assess your waste. You can draw your own sheet based on this and create other questions for this assessment based on your needs.

What Resources Are Available?

'Commuting sustainably is an area that I love. It is one that also benefits our pockets, the environment we live in and our health, while allowing people to mix, which, let's be real, our societies really need. These are benefits that we miss out on when we choose to drive alone in our AC cars, visit AC gyms and only dine at expensive restaurants. I'm not saying that we should never do this, but striking a balance will help us live a sustainable lifestyle on all fronts. Without this, you'll be missing out on the culture of public parks, the conversations or smiles shared on a bus, the fresh air under trees where families come together. I'm sure we don't need to be reminded of our planet's population, but there is so much diversity that comes with it. There are so many people from different locations, different backgrounds and so many stories to learn from that to not spend time with one another seems like a wasted resource!'

TRY THIS ACTIVITY OUT!

In this chapter, there are a number of suggestions that you can use for Activity #2. You can read ahead and return with new knowledge, or assess yourself now before learning a little more and return to this exercise later.

< 176 >

ASSESS YOUR
COMMUNITY RESOURCES:
ACTIVITY #2

Learn and/or take part in sustainability groups that aim to improve the environment in your community, such as climate action groups. Use the suggestions in the guide book or research to find more. Fill in each section of this activity to record your achievements.

The resources that I have learnt about are:

 I found out about online groups reducing waste in the ocean.

 I took part in a community clean-up.

 I researched about air pollution because I wanted to know how I could help.

Draw a picture from your activity to show friends and family:

The resources needed for this activity are:

 I used online searches to find out all I wanted to know.

 The clean-up organizers provided health and safety equipment.

 I visited a library, where I learnt about how communities can use air pollution monitors to align with climate goals.

Record where you learnt about these resources:

 The guide book inspired my search. I used online forums and social media to learn more.

Record who you shared your success with:

 My work colleagues! We took part in the clean-up together.

Use the above ideas to assess your resources and then create your own activity sheet.

Moving Towards a More Sustainable Lifestyle

'In India, we have amazing institutions that form communities of their own. They are where people come together. One of my favourite examples that best illustrates this is the Mumbai local trains. If you are on a train to Navi Mumbai or beyond, you'll find yourself becoming best friends with the woman next to you, who will buy and cut vegetables right there, and tell you what she is cooking for her family of five.[3, 4]

Another thing I love about Mumbai's local trains is how accessible they are, true to the welcoming ethos of the city. Although housing and many other services in India's biggest metropolis may now be unaffordable to many, travelling is still fairly easy thanks to the low fares. In fact, the Mumbai local fares are among the cheapest in the world, with commuters travelling distances of around 120 km for as little as Rs 30 (which is less than 50 cents in US dollars).[5] I would argue that Mumbai is best defined not by its skyscrapers but by its local trains and *dabbawalah*s (the city's famous lunchbox delivery and return system).

The dabbawalahs use the local trains to deliver home-cooked food to lakhs of Mumbaikars. They really need a whole book on themselves, with their entrepreneurial spirit and hustle, but for now these fun facts and a Harvard business case will have to do.[6, 7]

ZERO-WASTE COMMUNITY ORGANIZATIONS OF INDIA

There are many small businesses and organizations that are developing formal ways to support people and reduce waste in India. Listed below are some established organizations run by people who are passionate about their community and want to make a difference.

- **SWaCH**: Located in Pune, SWaCH is India's first wholly owned cooperative of self-employed waste collectors. It is an autonomous enterprise that provides front-end waste management services to the citizens of Pune. The cooperative covers over 70 per cent of the city, ensuring daily segregated waste collection from citizens' doorsteps while generating sustainable livelihoods for one of the poorest and most marginalized sections of Indian society. The user fee-based system has ensured high transparency, direct accountability to citizens and saved the Pune Municipal Corporation almost 100 crores in 2018.[8]
- **Saahas Zero Waste**: Located in Bengaluru, Saahas is a socio-environmental enterprise with over seventeen years of experience in waste management and resource recovery. It is a registered, empanelled vendor with the government in Bengaluru, Noida and Chennai to manage waste.[9]
- **Daily Dump**: Located in Bengaluru, it was 'started in 2006, [with a] vision to constantly re-imagine our relationship with the earth, with each other and with our urban spaces. Essentially, we are in the mindset changing business—mindsets about "waste", about marginal livelihoods, about

whose job it is to take care of "waste", about how we can harm less, etc.'[10]

There are other mentions in the 'Zero-Waste Library' section at the end of the chapter.

'I believe that the organizations listed above and others really want to help everyone in their community, in our societies, and make a positive change towards more sustainable lifestyles. This is a big plus, knowing that there are other people like this out there. Our future, green-thinking mindset literally needs to be, "I want to make any change I can."

Perhaps re-engineering our wasteful commuting styles, re-examining our relationship with our job or school or passions and rekindling a sense of community in these areas could help address some of the factors that hurt our health and the environment. We can start to make a change if we value the people around us and the resources that we have. Collectively, we can make a difference.

Think about how much cleaner, happier and stress-free we would all be with shorter commute times, cycling to work or walking to school. It is a mindset shift that can aid good health, higher levels of social interaction and reinvigorate the sense of community. Of course, there are larger changes involved (some of these are discussed in the appendix chapters of City and Travel), but let's start off with small simple steps: our everyday choices. Some of these actions are likely to have been the building blocks that those zero-waste community organizations used to establish themselves. Pretty cool when you think about it, isn't it? We're not talking about rocket science, after all!

So go over the next list of tips and tricks and study the illustrations that follow, which will bring to a close of the process that has been developing from

< 180 >

chapter to chapter (did you notice? If you missed the progression, flick back through the pages and see how interconnected the zero-waste life that you have been learning about is). When you begin incorporating them into your life, reflect on all of your learnings from the most private areas of your life to the most communal. Similarly, when you start studying the pictures, think about how the wasteful system followed a process from the first steps to the time it ends up in a landfill. Compare this with the earth-friendly version. I find it quite illuminating!'

ZERO–WASTE TIPS AND TRICKS

- Use public transport for your office, college, university or school commute,
- Sort waste (food waste, recyclables, toxic and rejects),
- Bring reusable containers for food, or invest in them for your office,
- Ask the office to change practices if they use single-use chai and coffee cups,
- Carry reusable items, e.g., water bottle, cutlery and straw,
- Use office materials made from recycled paper (including business cards),
- Change to fountain pens instead of using plastic ones,
- Join groups that promote sustainable practices such as e-waste recycling and valuing current technology instead of buying upgrades that lead to waste,
- Use hand dryers instead of paper towels,

- Install energy-efficient appliances, electronics and smart meters. You can have a whiteboard with the amount of water, energy and waste generated by the office on a daily, weekly or monthly basis. Get different departments, or offices, to compete on these metrics!

Why Is It Harmful?

'We're a long way from your personal care or closet choices, aren't we? Yet, every decision we make is linked to our community. A simple choice can reduce waste at your workplace, encourage a different way of studying at school, and maybe even put a smile on someone's face during a commute!

All of these lessons allow you to see why certain things that we are commonly doing today are harmful. We really should aim to become conscious consumers (and people in general) and begin to commute sustainably (or continue doing so) without wondering how big or small a change we are making. There are so many opportunities for us. In this section, we're going to discuss e-waste and managing resources responsibly, two big issues that more and more people are thinking about these days.'

COMMUNITY - CONSUMPTION TO LANDFILL

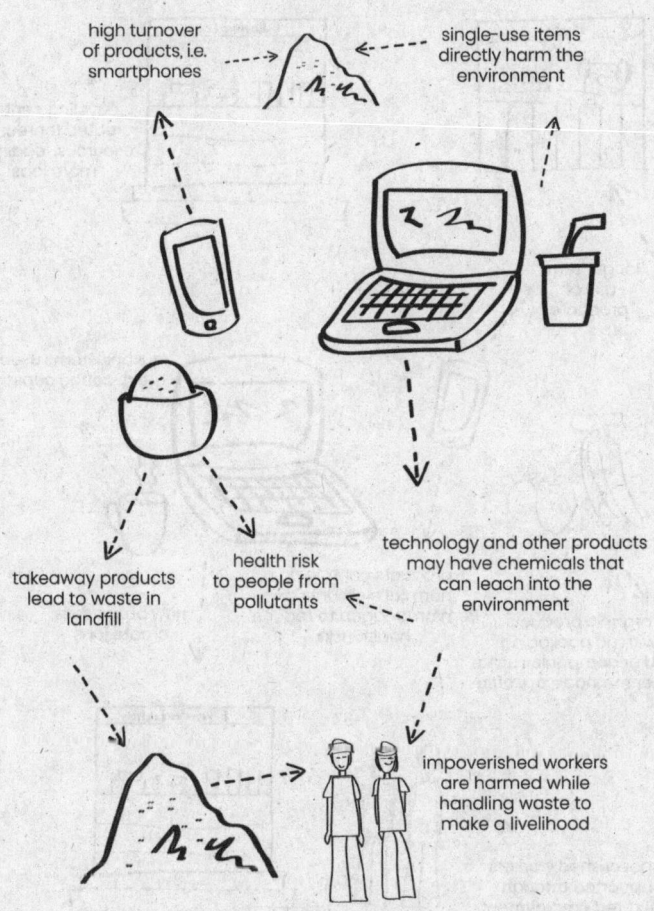

high turnover of products, i.e. smartphones

single-use items directly harm the environment

takeaway products lead to waste in landfill

health risk to people from pollutants

technology and other products may have chemicals that can leach into the environment

impoverished workers are harmed while handling waste to make a livelihood

COMMUNITY - CONSUMPTION TO REUSE

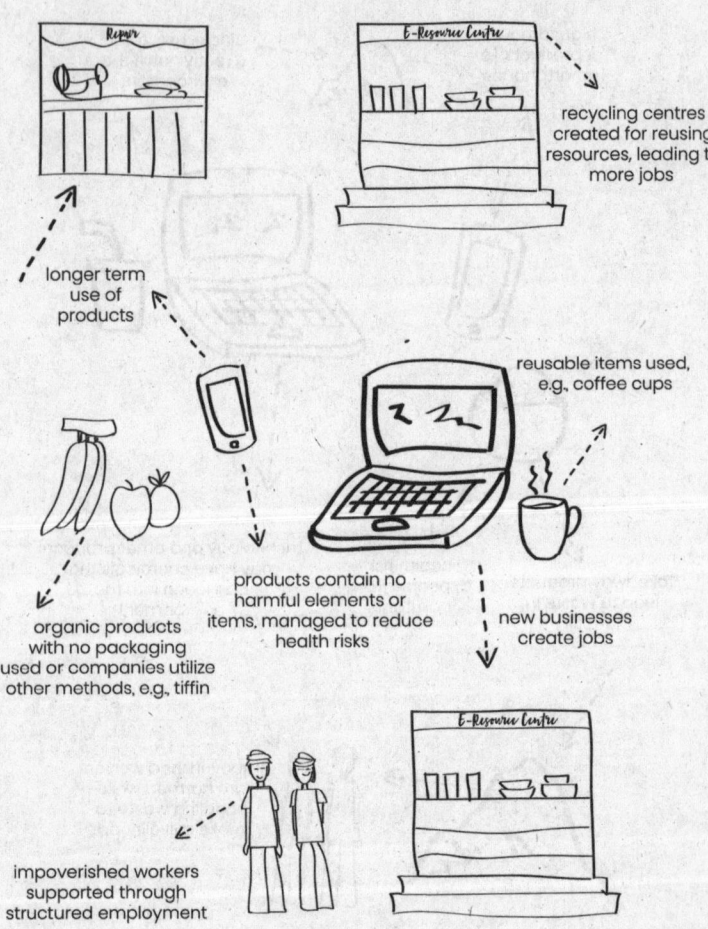

Repair

E-Resource Centre

recycling centres created for reusing resources, leading to more jobs

longer term use of products

reusable items used, e.g. coffee cups

products contain no harmful elements or items, managed to reduce health risks

new businesses create jobs

organic products with no packaging used or companies utilize other methods, e.g. tiffin

impoverished workers supported through structured employment

E-Resource Centre

You may spend your day in office, at school, or in a job that is set outside of a typical office environment, or in university or college. You may even be juggling a combination of working and studying. All of these areas, and many more, that you may encounter on a day-to-day basis (sports, music, theatre or dance groups to online gaming organizations and book clubs) are a part of your community. By being active within one or each of these areas, you can promote a waste-free society.

Conscious Consumerism

'Poor office waste management costs businesses an enormous sum every year and has a huge impact on the environment . . . on an average, 60–80 per cent of office waste produced by businesses is paper . . . each year around the world, 350 million printer cartridges are thrown away, having an enormous impact on the environment.'[11]

Many habits and practices that you may undertake on a regular day produce waste. This can range from buying takeaway orders at a cafeteria to the way that you use pens and paper. Many of the smaller areas that make tangible differences, such as reducing use of single-use plastic and becoming aware of practices that produce waste, have been discussed in earlier chapters.

Here, they will be mentioned briefly as positive examples that have already been implemented around the world.

- To reduce cafeteria waste, bring reusable containers, only buy what you need, avoid small packaged items, such as individual sauces, and ask the cafeteria to supply larger sustainable versions of these items (i.e. products in large glass jars). Additionally, recycle and segregate all items, encourage and/or implement composting practices. Adding signs to these areas to illustrate how these practices should be undertaken will help move your school/workplace towards waste-free operations.[12]

- Other ways to reduce waste include buying recycled products such as books, purchasing products with minimal packaging (avoiding packets with only one item is a good first step, finding alternatives without packaging can be the next), reusing older items for other purposes (as part of art-and-craft activities for younger school kids is a great way to teach environmental practices to youngsters) and keeping track of items such as pens so that there is no need to buy new ones due to lost items.[13]

- In an office specifically, you can reduce waste by saving your items on a hard drive instead of printing. Or, if you need to print, choose the double-sided option. If possible, use smaller font sizes to reduce ink usage. Using recycled paper is a good move, too. You can also reduce paper usage through avoiding paper bills by paying online. Additionally, use reusable items, such as paper clips, instead of staples and reuse envelopes and folders. Lastly, attempt to buy energy-efficient items such as light bulbs and other appliances. This will impact usage of both electricity and water.[14]

More importantly, in this chapter, the emphasis is on becoming aware of how all these factors are interconnected, both on an individual level and broader scale. Understanding how people, the planet and the entire waste system is interconnected will help you explain to the people you see and speak to why you are making changes.

Commuting Sustainably

In many areas across the globe, populations live close to major roads and highways and are thus exposed to tailpipe emissions from vehicle traffic, non-tailpipe emissions (such as brake and tyre wear), and noise that may have adverse effects on their health and well-being.'[15]

People who use public transport, walk or cycle to work or school have a reduced risk of health complications[16] than those who drive.[17] If you're asking why there is discussion about health in a book about waste, well, it too is a resource in the circular economy. One that can be valued and utilized to full capacity, or one that can be harmed. Becoming aware that you and the people in your community are resources that can be affected by practices, such as your commute, will enable you to view the interconnected nature of the systems in a new light.

Commuting sustainably can save an exponential amount of energy, reduce air pollution[18, 19] and alleviate traffic congestion and noise,[20] as well as limit the amount of waste created. Adopting new practices will limit the effects to the environment and to your health.

Well-informed decisions based on sound knowledge of an entire system have the potential to make a bigger change possible. They allow you to communicate with the people around you about why you are pursuing sustainable methods. A clear message about this will help to limit external pressure that you may receive in taking an alternate approach. For instance, if each of your colleagues commutes in a car and public transport is viewed as an unacceptable way to travel, by highlighting the reasons why you are choosing this method you may find that you receive less resistance. You may also find the others reflecting on their perceptions.

E-waste

India is the fifth-largest producer of electronics waste, or e-waste, in the world, generating close to 2 million metric tonnes in 2016. It faces a huge crisis with e-waste management. Although the Indian government introduced its first dedicated e-waste management policy in 2011 and expanded its scope in 2016, less than 5 per cent of e-waste in India is recycled through formally regulated units. The informal sector handles the rest, with very little control for environmental and worker health and safety.'[21]

Technology may be the biggest area of resistance to change because of the value attached to devices and global trends.[22] Although modern technology has brought tremendous innovation, there are numerous research articles highlighting the negative effects of how people function with these products in modern times and the environmental impact of the devices.[23, 24]

Information is one of the keys to a successful transition towards a zero-waste future. Of specific note, for the current e-waste discussion, is the United Nations Environment Programme's Executive Director who stated that the use of the linear economy in managing e-waste is 'an economic stupidity because [humanity is] throwing away an enormous amount of raw materials that are essentially re-usable . . . Whether it is gold, silver or some of the rare earths.'[25] The amount of 'materials that are available above ground in unused electronics now exceeds the amount still in the ground.'[26] Further, a 'tsunami of e-waste [is] rolling out over the world',[27] also known as the global e-waste flow. Estimates for 2018 suggest that there is 50 million tonnes of e-waste that has been produced.[28]

Also, there are extreme health risks associated with mismanaged e-waste, especially in countries with lower per capita incomes, where waste pickers collect and sort materials with limited to no protective equipment.[29, 30] The WHO notes this as an area of extreme risk, often due to 'primitive recycling techniques such as burning cables'[31] that can lead people to coming in contact with 'harmful materials such as lead, cadmium, chromium, [and] brominated flame retardants',[32] among others. Both human health and the environment are harmed this way. You, as a consumer and user of modern technology, can help. It is important to think about the effects of your discarded phone, laptop or other device on an individual level and on a broader scale.[33] In turn, this will allow you to make informed decisions about your next choice.

Globally, for environmental sustainability to be achieved, there needs to be a transition to looking at items such as modern technology as resources. Instead of e-waste, an e-resource could help achieve the needs of the planet. This is the circular economy.

Currently, there is a small but burgeoning number of stakeholders attempting to create a circular economy using modern-day technology. One terrific example is the first microfactory in the world created by Indian scientist Veena Sahajwalla based in Australia.[34] She decided to tackle e-waste because she saw clearly the issues these products can cause and the vast quantities that are produced.[35]

Sahajwalla's team understood that the reduction of costs of electronic products[36] has the potential to lead to more waste if facilities are not available to manage discarded items as a resource.[37] This growth in business is a fundamental component of a circular system, where jobs are created in new industries, such as recycling and repair centres. It also has the potential to formalize many sectors in need of more structured employment. Waste-picking communities could be a key beneficiary of such a transition, where formalized work allows for an improvement of livelihoods[38] and a more structured approach to managing the items that flow through a system.

Sharing knowledge about reduction in technology waste, be to alter the peer perception of a practice you are implementing to reduce waste, or simply as a topic of conversation over a casual coffee with an open-minded friend, is a good catalyst for change. Discussions of this sort may appear to be small and insignificant, but they hold potential to see more people becoming involved because of what they are learning, leading to a faster transition to a more sustainable system for all.[39] Notably, this discussion, although specifically related to e-waste, can be associated with many other areas previously written about, such as plastic and food waste, as the examples below highlight.

Responsibly Using Resources

'As far as animals go, we humans are pretty wasteful. And we tend to rely on the "out of sight, out of mind" model for most of our collective problems despite the fact that there's a fair amount of evidence that this does not work.'[40]

Communicating ideas has led to many encouraging signs in a move towards a circular economy, such as students at a school in the north-eastern Indian state of Assam who pay fees by depositing plastic,[41] or a juice vendor in the southern Indian city of Bengaluru who runs a zero-waste juice stall called Zero Waste Raja.[42] He sells his juice in fruit shells, such as watermelon shells, and uses banana leaf straws.

Some examples to reduce waste are:

- You can inspire your colleagues to change, e.g. ask them to bring in a container to use as a recycling bin. You can inform your colleagues about how to use it and what benefits it holds for the office and the planet.
- Organize a park, river or beach clean-up with your team.
- Provide unlimited filtered water and speak to people about why this is a better option than using single-use products.[43]
- Have reusable items in kitchens and canteens, i.e. plates, cutlery and cups.
- Reduce plastic in office chai and coffee by using loose leaves and beans instead.
- Encourage eco-friendly habits by gifting reusable products to the team, e.g., you can organize a Secret Santa event with a focus on sustainable and earth-friendly gifts.
- Ask your team for ideas on how to cut plastic use.

Additionally, if you're able to in your position, you can request that suppliers use less plastic in their packaging. It's also important to share your successes. This inspires others to act! Whether they're in another team at your school/workplace, or in another organization, it all adds to a larger community movement.[44] Moreover, encouraging and involving people in the process is a great way to make them feel important, which may lead to them being more involved, along with bringing new ideas and perspectives to the discussion.[45]

These are practical methods that illustrate positive lessons about how people, the environment and the waste system are interconnected, and how this connection can be improved through simple steps. Teaching people in the community around you, whether that is through signage on segregation or reducing plastic, implementing composting techniques, using recycled paper, or initiating large-scale environmental practices such as water- and energy-saving techniques are great strides forward that are being practised by many individuals in businesses, schools and social clubs around the globe. These methods can do a lot beyond reducing the amount of single-use items. They can create livelihood opportunities for people looking for better structured employment and can also increase awareness about other behaviours practised by people each day.[46]

At this point in your zero-waste journey, remember:

- Firstly, you are now more aware and knowledgeable of situations that you have control over on an individual level.
- Secondly, you have an understanding of the nature of systems and life cycles.
- Thirdly, you have been provided with other resources, such as recipes and checklists, which can be built into larger portfolios with additional researched information. Utilizing your new toolkit, whether at work, school, or elsewhere in your daily activities, will enable you to live a more earth-friendly life.

< 191 >

- Lastly, remember to think about the interconnectivity of situations around you. No matter what you choose to do, many other people in other parts of your town or city or further abroad will be impacted by your choices, just like you will be impacted by theirs.

What Can You See in Your Waste Now?

'Amazing work! We've made it to the end of our zero-waste journey, from our bathrooms to the world outside our homes. There are more resources for you to use at the end of this book if you want to look at the city-wide or global impacts of waste, but for now you have the tools to start your own journey towards becoming a zero-waste champion!

You can review your work and the entire process anytime, which is fantastic. As for me (aside from the bonus chapters at the end of this guidebook), I have one more thing to share about the future of waste in India. I'll meet you there.'

LOOK A LITTLE DEEPER WITH THIS ACTIVITY!

Activity #3 is a four-part process, on separate question sheets, that builds on everything you have learnt throughout the chapter. The importance of this exercise is to understand and become aware of the processes that can help you transition to a sustainable lifestyle.

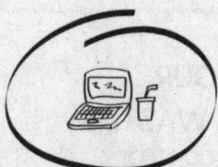

ASSESS YOUR COMMUNITY WASTE:

ACTIVITY #3 QUESTION 1 OF 4

What type of environmental impact do your products/services have?

Join the product/service to the problem using an arrow:

Product/Service:	**Problem:**
Car	E-waste
Stationery	Air pollution
Laptop	Landfill waste

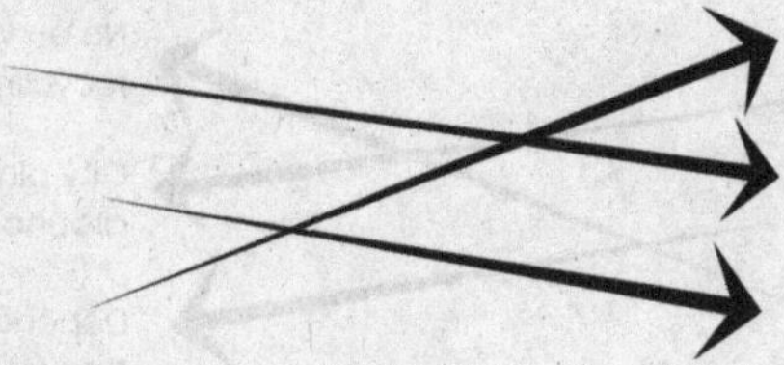

Use the ideas from this sample to assess your waste. The product/service may relate to more than one problem. Fill in the lines below with your products/services and environmental problems.

____________	____________
____________	____________
____________	____________
____________	____________
____________	____________

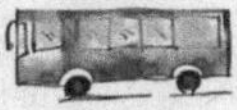

When thinking about the environmental impact, it is important to think of the number of areas that it could affect. Start on a small scale and work your way out. First, think about what it means to the micro-environment around you, then think larger and larger until you look at it from a global perspective. It will be beneficial to undertake some research online or in a library.

ASSESS YOUR
COMMUNITY WASTE:

What system issues prevent change?
Join the product/service to the issue using an arrow:

Product/Service: **Issue:**

Car   No e-waste
                                                          recycling centres

Stationery   City planned
                                                          around cars

Laptop   Dependency on
                                                          throwaway items

Use the ideas from this sample to assess your waste. The
product/service may relate to more than one issue. Fill in the lines below
with your products/services and environmental issues.

_____   _____

_____   _____

_____   _____

_____   _____

_____   _____

A system is anything associated with the product or service that is interconnected with it, for example, the
manufacturing unit that produces the product or the government office that provides the service. To learn
more, conduct research online or in a library.

ASSESS YOUR COMMUNITY WASTE:

What sustainable options are there to replace it?

Join the product/service to the solution using an arrow:

Product/Service: **Solution:**

Car   Find an e-waste recycler

Stationery   Rent or use public transport

Laptop   Invest in earth-friendly alternatives

Use the ideas from this sample to assess your waste. The product/service may relate to more than one solution. Fill in the lines below with your products/services and solutions.

_____   _____

_____   _____

_____   _____

_____   _____

_____   _____

A sustainable option is a product or service that will last longer and/or produce less waste. Think about options such as products made from earth-friendly materials, or those that reduce waste through a supply chain. To learn more, research online or in a library.

ASSESS YOUR COMMUNITY WASTE:

How will you start using the sustainable option?
Join the product/service to the action using an arrow:

Product/Service:

Car

Stationery

Laptop

Action:

Buy for long-term use and recycle

Buy an electric vehicle

Use sustainable options

Use the ideas from this sample to assess your waste. The product/service may relate to more than one action. Fill in the lines below with your products/services and actions.

_____   _____

_____   _____

_____   _____

_____   _____

_____   _____

This last step is all up to you. Make your choice, know the benefits and live a zero-waste lifestyle. You are more likely to succeed with support from your friends and family.

ZERO—WASTE LIBRARY

Here's a list of community-led organizations that can inspire. They help initiate and develop community collaboration.

- **Hasidurula**: This is a for-benefit, not-for-loss social enterprise that is focused on creating better livelihoods for waste pickers through inclusive businesses that have an environmental impact.[47]
- **Skrap**: It provides end-to-end waste management services for events. From identifying your specific needs to reporting the results of the waste management at your event, we help you to go zero waste.[48] Skrap has also managed waste for a series of music festivals like the NH-7 and Magnetic Fields.[49]
- **Y-East** (an initiative of Techno India Group): This is a platform that connects all individual and organizational actors of, from and for east and north-east India, basically all citizens, NGOs, corporations, investors, start-ups, public entities, educational institutions who truly care about social and environmental impact and want to do something about it in these regions, in their own desired and possible capacity. Y-East embodies the belief that cooperation and partnership is the key to greater positive impact, by gathering all these actors through one unique online platform, mechanically generating more offline opportunities to collaborate.[50]
- **Anthill Creations**: This is a not-for-profit organization that aims to 'bring back play for all age groups by building sustainable playscapes', using contextual designs and localized resources and encouraging community participation.[51]

There are many more like these. You can research online to identify more.

Community Waste Warriors Who Can Inspire

Listed below are some true community-level waste champions. They've built themselves, and their organizations, to a level where they make an impact within and beyond their individual communities.

- **Vani Murthy**: She's a composting enthusiast, urban farmer and waste management practitioner.[52]
- **Poonam Bir Kasturi**: She was trained as a designer but then moved on to found Daily Dump, an organization focused on composting and sustainable living throughout India.[53]
- **Nalini Shekar**: She is an Indian social activist and co-founder of Haisdurula.[54]
- **Manik Thapar**: He is the pioneer of the waste management industry in India. He is the founder of Eco Wise Waste Management, a leading waste management company in New Delhi.[55]
- **Dia Mirza**: She is a UN Goodwill Ambassador, UN secretary general's advocate of Sustainable Development Goals, the brand ambassador of Wild Life Trust of India, actress and producer who is on a mission to raise awareness about reducing plastic pollution.[56]
- **Sunita Nair**: She is director general of the Centre for Science and Environment. She has been with CSE since 1982 and has been heading it since 2000. Her research interests range from global democracy with a focus on climate change to the need for local democracy. She has also worked on forest-related resource management, water and waste issues.[57]
- **Nirupa Rao**: Nirupa is a botanical illustrator based in Bengaluru. Her illustrations are inspired by regular visits into the wild and are informed by close collaboration with botanists to achieve scientific accuracy.[58]
- **Pragya Kapoor**: The founder of Ek Sath Foundation (The Earth Foundation), her goal is to educate people and hold

them accountable for the effects their actions have on the environment. During the lockdown imposed due to the COVID-19 pandemic, Ek Saath Foundation supplied ration kits, as well as freshly cooked meals, to over 5000 people every day.[59]

- **Nayantara Jain**: She works with Reef Watch Marine Conservation, a non-profit organization focused on research, education and outreach, including beach clean-ups.[60]
- **Wilma Rodrigues**: She is a former journalist who now runs Saahas Zero Waste, an India-wide waste management company.[61]
- **C.B. Ramkumar**: Founder of The Sustainability Partnership, he is a sustainability expert, author, speaker, trainer and consultant. He is also an expert in helping corporates convert to total sustainability while factoring in global guidelines like the Sustainable Development Goals (SDG) of the United Nations. He is a board member and regional director for South Asia with the Global Sustainable Tourism Council (GSTC).[62, 63]

There are many more like them. You can research online to find out more.

E-Waste Recyclers

- **Bin Bag**: This organization provides dependable and professional e-waste recycling, asset recovery and data destruction services to business and organizations across India. Approved e-waste recyclers by the Pollution Control Board, they have two plants in Hindupur (Andhra Pradesh) and Guwahati (Assam).[64]
- **Karo Sambhav**: This is a country-wide organization currently spread across twenty-nine states, three union territories and over sixty cities. Through a technology-

< 199 >

enabled e-waste management programme, they provide producers and global brands with comprehensive Extended Producer Responsibility (EPR) services.[65]

- **Namo e-Waste:** They provide 'door-to-door services to ensure that your e-waste is collected with convenience' and transported to their recycling plants where they extract those parts which can be reused and sustainably dispose of the rest in an environment-friendly way. Their mission is to ensure that at least 50 per cent of e-waste produced in India is disposed sustainably.[66]

Lessons from Documentary Films

Listed below are a few movies that highlight community-led changes, aim to help the environment and raise awareness of problems within society. There are many more documentary films that you will be able to find with a little research.

- *Gasland* by Josh Fox: This is an exploration of the fracking petroleum extraction industry and the serious environmental consequences involved.[67]
- *Dark Waters* by Todd Haynes: This is about a corporate defence attorney who takes on an environmental lawsuit against a chemical company that exposes a lengthy history of pollution.[68]
- *The True Cost* by Andrew Morgan: This is a documentary film exploring the impact of fashion on people and the planet.[69]
- *An Inconvenient Truth* by Davis Guggenheim and Al Gore: It documents a former vice president raising public awareness about the dangers of global warming. He calls for immediate action to curb its destructive effects on the environment.[70]
- *Tomorrow* by Cyril Dion and Mélanie Laurent: It highlights climate change. Instead of showing the worst that can happen, this documentary focuses on the people suggesting solutions and their actions.[71]

- *2040* by Damon Gameau: This shows practical solutions to environmental concerns and hopes that the filmmaker's daughter, a twenty-one-year-old in 2040, will face a hopeful future.[72]

How to Set Up a Community Garden[73]

Community gardens can help us with one another and nature, thereby making us feel more rooted and accountable to looking after the planet. The other benefits they offer are:

- Assist in improving food security,
- Enhance a sense of community,
- Reduce carbon miles (emissions) of your own food,
- Encourage the growth of a sharing economy.

To create your own community garden:

- Start a conversation in your community to gauge interest levels,
- Find a small piece of land or an area where you are allowed to set up your garden (check with local officials if you are unsure),
- Put together finances for the project. This can come from you, a community group or other organizations. Ensure that the funds are available and that they are enough before you begin,
- Gather a group of engaged community members to assist with the planting and management of the area. Make sure that the people involved are able to water and maintain the area,
- Allocate specific areas for individual plants,
- Create a composting area that can be used as a resource to nourish your community garden,
- Collect the seedlings or infant plants together. This can be done by individual participants bringing their own, or

< 201 >

the funds gathered being used to purchase all of the items you need,
- Build a fence to protect the area,
- Organize a watering and management schedule that is in line with people's availability (you probably know their availability due to the information you gained when recruiting team members),
- Once the garden is set up, you can use it for community meetings, events, presentations and other community get-togethers. This area is one that your community created. By sharing it, having meals surrounded by plants and enjoying your time there, both you and your team members will have achieved the goals of a community garden,
- Enjoy it!

Sweet Sweat
A DIY trick to deodorize your shoes or camping bag.

You will need:

- 2 tablespoons baking soda.

To naturally deodorize your shoes (or gym bag), all you need to do is:

- Sprinkle the baking soda inside,
- Leave it for 5–10 minutes,
- Clean the powder out.

This will clear all the smell away (if it's still smelly, try it again. Leave the baking soda in for longer).

Stain, Stain, Go Away (Naturally)!

This is a DIY way to make a stain remover.

You will need:

- 4 tablespoons baking soda,
- ¼ cup water.

This method is great for any stains on your clothes. All you need to do is:

- Add baking soda to water in a mixing bowl or bucket,
- Mix well till it becomes a thick paste,
- Soak your clothes for approximately 3 hours,
- Wash as usual after this.

This natural method effectively removes perspiration stains, rust stains and even fresh grease!

Smelly Feet? It's Not Your Fault!

This is a DIY foot soak for smelly feet.

You will need:

- 1 cup apple cider vinegar,
- 4 cups of warm water.

To ease the smell away from your feet:

- Add apple cider vinegar to a bowl, followed by warm water,
- Stir this mixture,
- Soak your feet for 20–30 minutes.

The antimicrobial properties of apple cider vinegar helps deodorize your feet, while the antifungal properties can help combat and prevent conditions such as athlete's foot!

become mindful about the type of material that your personal care products are made from or packaged in

choose to promote sustainable choices in your community, like using recycled paper

ensure that you compost wet waste in the kitchen. Avoid plastic

ZERO-WASTE LIFESTYLE

shop local and commute on the ground while you travel

limit your impact on the environment with your gifting choices

begin using organic products throughout your home to maintain a good level of home care

think about your fashion choices, which can include knowing the entire supply chain of a product that you keep in your closet

become an active citizen of your city by using public transport to commute

CHAPTER 7

TRANSITIONING TO A ZERO—WASTE LIFESTYLE: THE CIRCULAR ECONOMY AND SUSTAINABILITY

Until now, the book has discussed a range of issues concerning waste, both large and small. It introduced the impacts of plastic waste in Chapter 1: Personal Care, while Chapter 2: Closet offered insights into the supply chain, material waste and a small discussion on plastic. Organic waste and how systems are managed broadly was spoken of first in Chapter 3: Kitchen and expanded upon later, as was the idea of waste from packaging. Chapter 4: Home Care provided a more in-depth view of chemicals and sustainable practices at home, while Chapter 5: Gifting discussed packaging waste more and outlined the impacts of pollution (that of air, for instance) on health. Throughout all the chapters, but specifically in Chapter 6: Community, the point of what a resource is and how it should be used was presented, along with waste from technology, commuting and the effects of our everyday practices. If you decide to read the two concluding chapters, which discuss wide-scale effects, Chapter 8: City and Chapter 9: Travel look at the impact of waste on humans and the environment. The former discusses areas associated with sustainable cities, while the latter looks at solutions for the issues that have arisen due to waste building up around the world.

This guidebook has hopefully provided you with insights that will enable you to make well-informed choices from a highly educated perspective, particularly if you undertake further research or try the many DIY activities and recipes in the 'Zero-Waste Library' sections of each chapter.

Things to Keep In Mind

- **Systems**: There are two main systems. They can be complex or simple depending on how you utilize them. Understanding the reasons why a linear system produces waste will enable you to articulate to others, and to reassure yourself, why a circular economy is beneficial to all.
- **Business As Usual**: Change may often seem to be slow. This is partly because of the current reliance on systems

of production (which are dependent on fossil fuels, for example). The fact that business occurs in this fashion should be neither surprising nor discouraging. Try and evaluate what changes have happened and identify the areas that have not changed in global production and consumption patterns as an opportunity for the future rather than a failure.

- **Timelines**: It is important to understand that many wasteful practices worldwide have been developing for over two centuries. There is a need to change, but there needs to be a realistic understanding of how it will with set timelines and goals that are provided to all in global frameworks, such as the Sustainable Development Goals.

- **Optimism and Pessimism**: The peaks and troughs that you go through on your journey towards sustainability should be expected. There will be times when things get overwhelming and others where you make great headway. Try to maintain a balance and stay objective. This, in turn, will help you be part of a collective (worldwide) problem-solving and strategic-thinking transition (note that this is not an overnight move) away from a linear system and towards a circular system.

- **Waste**: Understand why the waste is there, how it is viewed, why it is not being utilized and other such issues. By understanding these on an individual level, you will be able to make tangible changes in the immediate sense. Understanding these more broadly will enable you to be part of a global movement for the future.

- **Impact**: Impacts are both positive and negative. Choosing to aim for positive goals is important, but it is also vital to know why and how the negative impacts are happening. Through an understanding of these factors you will become more capable of making positive changes, no matter who you are or where you are in the world.

- **Resources**: Perception and a world view is paramount when it comes to transitioning to a zero-waste lifestyle.

It can only be achieved if everything becomes a resource. Small steps towards this can be made in the immediate sense, while broader impacts made in a city or globally will require collective co-operation and efforts with all stakeholders being transparent about the components that go into the functionality of whatever they are doing. With this, everything has the opportunity to become a resource, thereby allowing everyone to achieve a zero-waste lifestyle.

What's Next for Sustainability in India?

'For the first time, there is growing consensus about the environmental challenges that we face. There are growing voices and a growing awareness. There is a development of methods and techniques. There are numerous people who want to be involved. Every single person is crucial. From coconut vendors who refuse to use straws made of plastic to student action groups focused on climate change and air pollution in our congested cities and the waste pickers I see in my neighbourhood, who understand the realities of wasteful practices. Everyone matters in a move to local and global sustainable methods.

Yes, it is late. Yes, we are feeling the effects of plastic pollution. Yes, there is air pollution. Ground pollution. Water pollution. Higher levels of extinction than at any other point in human history. Natural environments have been destroyed. Some are on fire. Some are inundated with rain. Lands are becoming less arable for the crops that we need to grow to feed a booming population. Large areas of the planet are completely infertile. Completely beyond a state of repair. Icecaps and glaciers are melting, too. Yes, we know this.

But what we also know is that small changes can make tangible differences. That can be achieved by making your own soap, by refusing to use single-use products, by choosing to compost your waste, by taking public transport, by sitting down and taking your time over a cup of chai or coffee instead of a to-go beverage in a single-use cup, by talking to

< 208 >

people about sustainable issues, by spending a day in an affected environment to become more aware of the situation, by hearing from people whose lives are closest to the threshold, by being empathetic to the people who feel the impact most profoundly, by listening, learning and making decisions with all of these knowledge sources.

We must ensure that we do not become too pessimistic. Neither should we be too optimistic and lose sight of the gravity of the situation.

We're a land of amazing contrasts, colours, traditions, sights and smells. We have an opportunity to utilize it. We can learn from others. We can share our own knowledge. We can become a catalyst for change simply by becoming involved with simple actions. We can do even more by involving ourselves in broader conversations.

There are opportunities for change in India. The solutions will most likely be specific to our country. That may be really important and successful compared to if we try to directly replicate other countries. We can learn from external locations, of course. Yet, our uniqueness will require us to modify based on how we live. We can combine our traditional sustainable methods, which worked for centuries, on a small scale with new technology that allows these techniques to be used for a growing society. Or maybe it's time for new innovations, like many already have. We can utilize our collective minds.

There are over a billion of us here in this country. We are a land of entrepreneurs, which also means that we are the land of new, bright and curious insights and ideas. We can utilize this resource like few other locations in the world can. It is likely that numerous solutions will be found that will develop into multi-tiered approaches. Some that the government will help with due to the large budgets needed, some that large and small businesses can become involved in, while others may be achieved on individual levels through individual action and choice.

Today, we are at a point where people, profits and the planet can be valued equally. For a long time, the middle 'p' was the focus. For a circular economy to be effective, all three need to be valued. This holds true even more now, when terms such as sustainability, circular economy, zero waste and valuing resources are becoming the rhetoric. This is a huge positive. A time of conscious consumers, active citizens, people taking responsibility, and, perhaps, the initial signs of a country ready to change its ways.

For all of us, we can see waste everywhere, we recognize the effects, we want change and know how to achieve it. Truly, there is no better time to start valuing everything as a resource than now.'

This Is the Perfect Life

You wake up in the morning. You shower using water that comes from the collectively managed supply, which in turn comes from a local source and is supported by many independent groups and run by the elected government, like many other facilities these days.

The natural soaps and personal-care products leave behind a lovely earthy smell. You made these in the local co-operative centre that everyone in your community visits. The centre is supported by a group of women who are making these products using the recipes handed down to them from the previous generations and the technology available today.

Once you have dried off, you move to your closet and choose your clothes carefully. You liked all the options at the clothes swap the previous night. Now it is as though you have too much choice until the next clothes swap you visit, which will be the next month. You love how clean the items are each month and how the new apparel provides a unique angle and a story. Stories, as you know, are important.

All products are rented and repaired these days, from your phone to your washing machine. Anything and everything. The method supports the growing workforce. The system has helped everyone in the city live a healthier and more mindful life.

Your home is powered by renewable energy that is managed collectively like the water supply. You think about how lovely it is that everyone is open about what they do and how they do it. It has helped so many people live a better life.

You smile as your compost is collected to nourish the farms on the roofs in the centre of the city. The city has been cooler and greener since the government acted on everyone's desire to create a more sustainable environment. The local action groups explained everything clearly and provided the catalyst that the government needed to know that they could make the monumental change.

Now, the power and resources move in and out of the urban centre easily due to the transport network that connects rural locations, satellite towns and the city in no time at all. All systems balance each other out and function as part of a cohesive whole. You love the interconnectedness of it all.

You have decided to take the train and head out to a remote beach today with your friends. There are so many fish in the ocean again, just as there are forests that have been regrown, with an abundance of wildlife flourishing within, which the train will skirt alongside for the ride.

So much has changed. Joyfully, you realize that these thoughts are only a small glimpse of the whole. You smile to yourself, believing that the day will be remarkable, just like everything else around you has been ever since humanity decided to make a collective choice.

take your waste home if there are no correct disposal options

help groups and initiatives that support sustainable practices, such as green cities

ZERO-WASTE CITY

vote for politicians focused on sustainability

use a reusable coffee cup

commute sustainably

VOTE

always carry reusable bags and containers

take your time, e.g., sit down in a cafe or restaurant

CHAPTER 8

CITY

'The effectiveness of green technologies in modern cities, especially in waste management, depends on the level of participation of citizens. People are active participants in the life processes of cities and have a direct impact on the urban environment.'[1]

The Perfect Life . . .?

You are travelling through your city. There is a red car beside you and a blue car in front. On the other side is a truck and behind that is a bus. The noise from the vehicles is drowned out to a certain degree by the radio, which is turned up to ten. Your lunch with your friends had to be cut short because all of you were in a rush to get to your other engagements. The takeaway packet containing your half-eaten meal sits on the empty seat beside you.

The light turns green. The blue car in front releases the brakes. The truck chugs into gear. The red car stalls for a moment. The bus beeps at you as you speed over the black bitumen.

You see your city standing tall around you. With just your takeaway bag for company, you roll past parks that are empty because of the soaring heat. The cinema next to one of the concrete playgrounds has a long line outside it. People are pushing in to try and get some cool air on to their faces.

You turn the corner. Then the next. You climb the small bridge towards your neighbourhood and pay little attention to the river that has never been itself since the large company started production upstream. The sun is covered by a grey haze. You wanted the government to address the smog issue, but they never seem to do what they promise before they are voted in. Hardly anyone expects a change any more.

The road that your car is bumping over has a lot of people living on the side these days. They hide under the shadow of the new apartment buildings and restaurants. This reminds you to make reservations for dinner with your colleagues, family and friends next month. You know how lucky you are to live within driving distance of these fine restaurants. Other people you know may need to take public transport to get to the restaurant, in part because of the limited car spaces available, but not you. You plan to get there early so that you can park your car.

You take the next left, the right after that and pull into the driveway. That's it! Just two blocks from the restaurants. You pick up your phone and dial. You plan to set the dates for your next dinner so that you do not forget. You walk to the bins that are overflowing a little, place the takeaway lunch that you no longer want into one of them and shake your head at the poor garbage collection in this city. On the upside though, you feel, the city has nice restaurants and many other areas that you like to frequent. Someone at the restaurant picks up the phone and you make a reservation.

What Is in Your Waste?

'Bengaluru, the IT capital of India, is infamous for its garbage crisis and holds the title of "garbage city of India". A city with a population of 1.23 crore (more than 12 million people) generates 4000 metric tonnes of waste on a daily basis.'[2]

'More and more people are living in cities these days. They frame the contexts of our lives, which is why they're probably the most complex topic in this book. I think it's refreshing to think about citizen engagement and what that can do. Every little bit and every person within a city combines to make a more sustainable world. Assessing these areas will allow us to see how things can change on both a large and small scale.

I'll start with a short discussion about Bengaluru's steel flyover before discussing the now-famous waste clean-up in Mumbai. These are two of India's largest cities. If a change can be made here, it can be made everywhere. Don't you think?

Bengaluru is notorious for its traffic congestion. It is a city that takes a good portion of the day to get from one side to the other. This will occur no matter what time of the day you leave, unless you are fortunate enough to live near the metro line. This relatively new mode of public transport has really shaken up how people in the city view commuting. It has led to a lot of positive experiences, including shared spaces, and has in turn helped ease congestion and reduce air pollution.

But back to the flyover. A proposed 6.7 km-long steel flyover was recommended by the government as a way to ease traffic congestion in the north of the city. The Rs 1761 crore-proposal[3] suggested that the flyover would not only ease traffic but also require around 800 old, magnificent trees to be cut down.

A couple of incredible events were triggered by this. Firstly, independent research, conducted by the Indian Institute of Science, highlighted that the proposed flyover would not ease congestion substantially. It was found that the flyover would reduce Bengaluru's tree cover by almost 3 per cent.[4] That's an incredible amount for a single project in a city as large as Bengaluru. The institute's recommendation was that more money and resources be placed into expanding the metro line. Secondly, over 8000 citizens came together to protest the destruction of the trees, forming a giant human chain along the proposed flyover route towards the airport. This, to me, was an excellent example of citizen engagement. People from across the city, geographies, socioeconomic backgrounds and ages came together. The human chain involved six-year-olds and even well-known historians such as Ramachandra Guha.

These factors spurred community groups into action, including many spokespersons who emphasized the main reason behind why governments focus on development through roads—corruption and financial gains for certain groups of people. The collective action and clear, precise and well-researched facts communicated to the elected representatives in an effective and ethical manner prevented a mass tree massacre. Accurate data, facts and citizens' involvement changed the course of the city's development and in turn aided in maintaining the trees that effectively combat air pollution.

This highlights the immediate need for all of us to come together and have our voices heard, be it through peaceful protests, through your job, or through those you meet on a daily basis. Considering the magnitude of the challenge we face as a community, with climate change, inequality of all sorts, corruption and the possibility of financial gains taking precedence over everything, even after community voices might be heard. It just means we need to become that much more cohesive, consistent and aware of all that's happening around us, and continue striving to have our voices heard.

The Bengaluru example is a great sign of what citizen involvement can do when governments seek immediate financial incentives over long-term environmental sustainability. However, what happens when there is already a mess? One of Mumbai's  beaches has become the classic example of citizen involvement. Versova beach was covered in waste that had washed up from the ocean and because of poor waste disposal.

One day, a local lawyer named Afroz Shah began picking up the waste from the beach. As time went by, more people joined. As the group became larger, the beach became cleaner. Government and businesses began to notice. It has now become a huge environmental movement, bringing together people from all walks of life, from celebrities to faluda stallwallahs to students.

Of all the signs of progress, I like this one the best. For a twenty-year period before the waste was acted upon, or even noticed, the turtles that once inhabited the beach were no longer existent. It was desolate. However, once the beach was cleaned, and the state of Maharashtra, Mumbai in particular, began implementing stringent waste management plans, the beautiful turtles who had once lost their homes to piles of waste returned.[5] Soon, not one or two, but eighty Olive Ridley hatchlings were waddling around on the beach for the first time in twenty years! Other major wins included the fact that once the initiative began gaining traction, local authorities began contributing amenities and machinery to help accelerate the process. The United Nations even declared this the world's largest beach clean-up (5000 tonnes of waste was cleaned up, often found up to 5 ft/1.5 m deep in some areas)![6]

Of course, there are many factors that go into this. However, it is a sure sign of how nature can rebuild if citizens are actively involved and if governments are influenced to implement long-term environmentally and socially sustainable policies. Change can happen because of simple acts that can help turn an unsustainable, waste-filled tide into a smart, sustainable city. We underestimate how many people are watching our actions, what we say and how kind or unkind we are to each other. If we realize this and share our ideas, we will be able to influence each other simply through actions rather than forceful arguments or conflicts.'

NOW IT'S YOUR TURN TO ASSESS!

Try out Activity #1 with your current knowledge. View this activity as an initial way to see how much you know. Once you have learnt more as we go forward in this chapter, you can return to it in order to see how much your knowledge has developed.

< 218 >

ASSESS YOUR CITY WASTE:
ACTIVITY #1

Follow the example sheet shown by choosing a few products/services (for services it could be something like transport) you use to assess.

	Example A	Example B	Example C
Is this the best form of energy available?			✓
Is the product/service harmful to the environment?			
Does the government involve the citizens?			
Is the water source the most sustainable option available?			

- If yes: share the solutions with friends and family.
- If no: research options.

Use the ideas from this sample to assess your waste. You can draw your own sheet based on this and create other questions for this assessment based on your needs.

What Resources Are Available?

'Another example that I really think shows how people can make a difference is in the state that I live in. A few years ago, the government of Karnataka introduced a competition for creating master plans for various cities.[7] It brought together people from all walks of life to help create smart, sustainable and future-focused options. The entries were really well received. The process was extremely democratic, allowing the government to listen to the people. It worked in tandem with the funding and resources that governments have access to. It also involved the Bruhat Bengaluru Mahanagara Palike (BBMP), the local administrative authority responsible for civil amenities, in the creation and implementation process.

This step of involving people across the board and really valuing them can be a big step forward as far as sustainable cities are concerned. The way I see it is that the average citizen cannot make an entire city more sustainable. You and I cannot build a metro line or implement large scale single-use plastic bans for an urban centre independently. It shouldn't be that way either. In a democratic country such as India, the government should not implement projects such as the proposed steel flyover without consulting the people whose lives will be affected.

Involving all the stakeholders can help ease the transition towards making cities more sustainable. It will require large government budgets and commitments, but it will need something that we already have in India, i.e. active involvement from individuals, small businesses and larger companies. There are numerous people who have already started to bring a change and ensured that the government takes notice.'

TRY THIS ACTIVITY OUT!

In this chapter, there are a number of suggestions that you can use for Activity #2. You can read ahead and return with new knowledge, or assess yourself now before learning a little more and return to this exercise later.

©Bare Necessities Zero Waste Solutions Pvt Ltd

ASSESS YOUR
CITY RESOURCES:
ACTIVITY #2

Use sustainable transport for a daily activity, like work or shopping, and/or several trips over the course of a week. Use the suggestions in the guide book or research to find more. Fill in each section of this activity to record your achievements.

The resources that I have learnt about are:

 I cycled to university on a new bicycle route.

 I walked to the shops to collect groceries, instead of driving.

 I took the new Metro to work.

Draw a picture from your activity to show friends and family:

The resources needed for this activity are:

 I used a rental bike, from a bike-sharing service in the city.

 I took my reusable bags.

 I learnt about the Metro route and used it. I'm going to continue using it. It's cheaper, faster and cleaner than a car.

Record where you learnt about these resources:

 The guide book's 'Zero-Waste Recipe' section.

Record who you shared your success with:

 My colleagues! They are now all using the Metro, it's a big positive change.

Use the above ideas to assess your resources and then create your own activity sheet.

Moving Towards a More Sustainable Lifestyle

'Active citizen involvement is key to driving change. It is something that both you and I can do as a starting point to encourage large-scale change. A lot of the time, we think such change starts with government regulations. While a lot of it does, we must understand that governments have limitations, too.

The change away from a linear business model can be achieved if environmental action groups sit down with elected officials and explain the issues clearly to them. This is what happened in the case of the Bengaluru steel flyover. Communication produces many different ideas which can address the waste crisis. No matter what the method, the fact is that city-wide issues need to be implemented by those with the budgets and the ability, i.e. the government. Small changes can be made by citizens, as catalysts for bigger change. Large systemic transitions, however, can come through active government involvement. Clear communication and accurate data are highly useful in making good decisions that will benefit our future urban centres.

It is important to not become oblivious to issues. If we walk past a pile of waste, or see traffic congestion, we notice it one day. The next day, we see it again. By the third day, it becomes the new normal. Therefore, it is important to be aware of the issues and communicate in the right way. Given the large-scale issues we are facing today, it is important for us to understand who we are communicating with and their respective journeys, so that we can communicate in a manner that is acceptable to a specific person or group. We must involve everyone and be aware of the problems in order to find solutions. Every conversation that increases awareness helps.

One of my favourite examples is Slovakia's first female President, Zuzana Caputova, who is now the youngest person to serve as the President of Slovakia. She is sometimes called the Erin Brokovich of Slovakia for her decade-long struggle to close a toxic landfill in her hometown, which she succeeded in doing. President Caputova was a public interest lawyer and active citizen for over a decade before she got involved in politics and governance.'[8]

ZERO—WASTE CHAMPIONS OF INDIA

Below are some inspirational Indian enterprises who have shifted the way that sustainability is viewed in cities. They may not have the title that the Slovakian President has, but they are just as important in ensuring a waste-free country.

Research a little more to find out more about this quartet and others who help reduce waste in various ways.

Sustainable solutions from businesses and individuals are having impacts in various ways, from coconut vendors who use coconut-leaf straws instead of plastic to waste pickers gaining more structured employment support. Additionally, innovations from zero-waste champions have an impact at the national and international levels, involving prominent people in the government who are in positions to promote these solutions.

* **Evlogia Eco Care** (creators of the coconut leaf 'Leafy Straw'): This is a start-up that comes up with eco-friendly and healthy innovations meant for daily use by common people. Tapping into the knowledge, expertise and creativity of people who believe in working towards a sustainable world, Evlogia has already come up with the

patented 'Leafy Straws', an alternative to plastic straws, made entirely out of fallen palm leaves.[9]

- **Kabadiwalla Connect**: They help leverage a city's existing informal waste infrastructure in the collection and processing of post-consumer waste. Using ICT- and IoT-based technology, they integrate informal actors into the formal waste management system to deliver cost-effective and low-carbon waste management solutions that cities in the Global South need to support their growing economies and populations.[10]

- **Paperman**: This was started in 2010 with an objective to accelerate recycling in India. Their initiatives include public awareness programmes, on-demand recycling platform, managing processing units in partnership with various state governments, turnkey contractors for setting up recycling units and other ancillary services. They provide world-class technology to clients, partners and public organizations to design and execute large-scale recycling projects.[11]

- **Stree Mukti**: This organization has been working towards women empowerment for over four decades. The play *Mulgi Zali Ho* (Girl Is Born) opened doors for women to interact and share their problems with Stree Mukti. The family counselling centres, in-house monthly publication *Prerak Lalkari*, programmes for adolescents, day-care centres, programmes for waste pickers and solid waste management have enabled sustainable livelihoods for women.[12]

'These are terrific role models for sure! However, not everyone can be a big influencer or start a business that has the opportunity to bring about city-wide changes. In fact, many don't want to be, which is perfectly fine! Remember that every little effort, added with others across the city, can make a monumental difference. You can move towards a zero-waste lifestyle in numerous, easy-to-achieve ways.'

ZERO—WASTE TIPS AND TRICKS

- Commute sustainably, i.e. walk, cycle, carpool or use public transport,
- Always carry reusable containers and bags with you,
- Use a reusable coffee cup,
- Have an ice cream in a cone rather than a plastic cup,
- Take your time, e.g., sit down and eat in a café or restaurant,
- Take your waste home for correct disposal if there are no options available where you are,
- Support groups and initiatives that promote sustainable practices, such as green cities,
- Vote for politicians focused on sustainability.

'Cities all across the world can do one sustainable thing to improve quality of life: invest in shared transport such as Metro lines and public buses. In India, we can learn from cities that promote sustainable transport in Europe, such as Copenhagen and Amsterdam.[13] Other urban centres can then try and replicate aspects of their sustainable processes in ways that work within their environment. Or we could look at systems already implemented in some of India's most sustainable cities, Pune and Indore.[14]

Another stride in an environmentally friendly direction is to encourage businesses to employ people who live near the office, so that commutes are reduced or can be done sustainably. Walking to work, for instance, is being promoted in cities such as Melbourne, a concept called "twenty-minute neighbourhoods",[15] where people are encouraged to live within a short distance from their office or school. New York is another major hub that values pedestrians and other forms of sustainable transport. India can learn from these templates and implement it in ways that function well for it.

Additionally, "satellite towns", where economic hubs or places of work are located around a region instead of solely in the central business district of the city, may diversify the flow of migration to other urban centres given the economic opportunities there.

It is complex, but as a species we have accomplished so much through knowledge acquired using the resources available to us at a specific time. These days we have so much more knowledge and access to resources. There is a lot of scope for entrepreneurship and innovation to help rebuild cities and systems into more sustainable ones. We can choose to move to a people-centred way of living. It is likely that this will be a method that is embraced by every citizen who is a part of the total system of the city.'

LINEAR CITY SCALE IMPACTS

road congestion and pollution because there are limited sustainable transport options

leads to climate change due to carbon emissions

waste fills dump sites, harmful for people and planet

creating heat centres

overcrowded areas

people are at health risk

unsupported waste workers

overpopulation and land clearing, due to poor planning

polluted waterways that are a key water source

CIRCULAR CITY SCALE SOLUTIONS

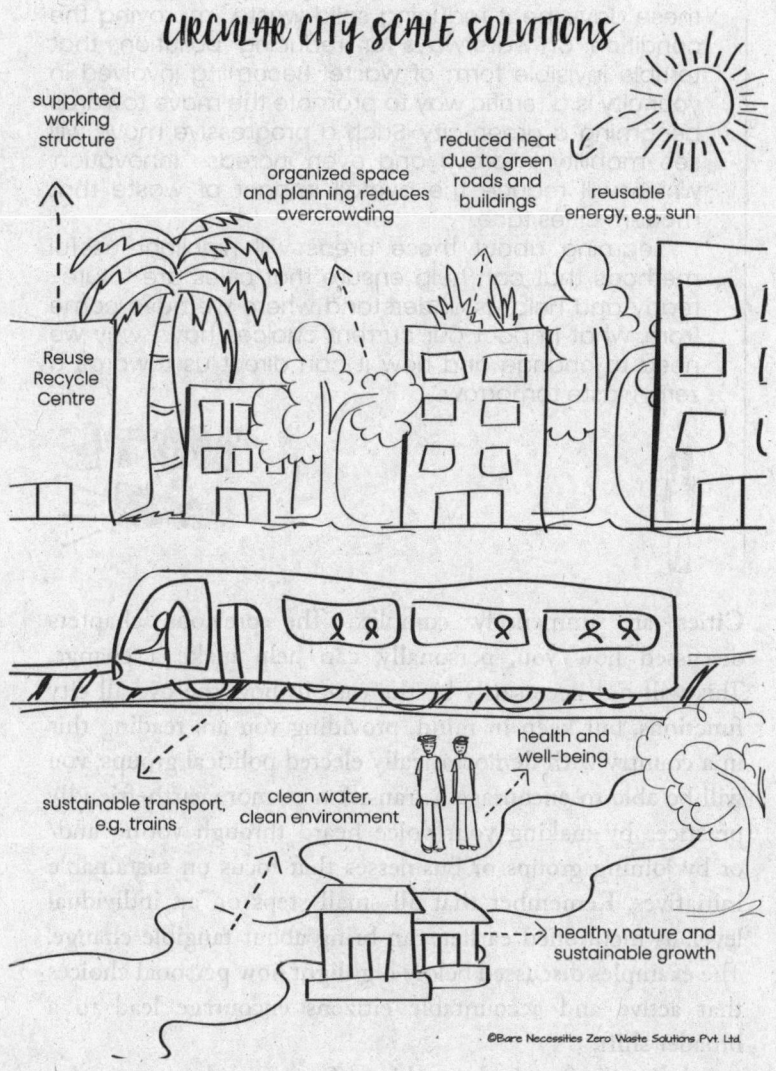

supported working structure

organized space and planning reduces overcrowding

reduced heat due to green roads and buildings

reusable energy, e.g., sun

Reuse Recycle Centre

sustainable transport, e.g., trains

clean water, clean environment

health and well-being

healthy nature and sustainable growth

©Bare Necessities Zero Waste Solutions Pvt. Ltd.

Why Is It Harmful?

'There are so many exciting things happening in cities these days, be it reducing solid waste, improving the condition of waterways or reducing pollution, that terrible invisible form of waste! Becoming involved in your city is a terrific way to promote the move towards becoming a green city. Such a progressive move will see mobility improve and even increase innovation, which will reduce the overall impact of waste that modern cities face.

Learning about these areas will highlight useful methods that can help ensure that cities are future-ready and help us understand where we have come from, what impact our current choices have, why we need to change and how it can direct us towards a zero-waste tomorrow.'

Cities are immensely complex. The previous chapters discussed how you, personally, can help make a change. This will not necessarily be the case in how the overall city functions, but keep in mind, providing you are reading this in a country with democratically elected political groups, you will be able to encourage a transition to more earth-friendly practices by making your voice heard through voting and/ or by joining groups or businesses that focus on sustainable initiatives. Remember that all small steps on an individual level, as mentioned earlier, can bring about tangible change. The examples discussed below highlight how personal choices that active and accountable citizens encourage lead to a broader shift.

We will now look at things from a wider perspective compared to the previous chapters.[16] Our view will now focus on wider areas such as improvements in mobility in the city[17] and green/smart cities reducing the heat island effect.[18]

Mobility in the City

'Improving urban mobility is crucial to the evolution of every city, but it is arguably even more fundamental for the future of evolving megacities.'[19]

The ability to move within a city is a key human need and is often viewed as an enabler of both economic and social prosperity.[20] Yet, with an increase in population and the trend of migration from rural to urban areas[21, 22, 23] mobility is hindered. Unplanned expansion leads to traffic jams thanks to more vehicles being on the roads.[24]

Research from the USA has found that restriction of mobility cost drivers $87 billion (USD) in 2018, which means $1348 (USD) per driver.[25] The issue, however, is not restricted to the USA. The top five most congested cities worldwide[26] are in Russia, Turkey, Colombia, Mexico and Brazil.[27] Along with monetary loss, the increase in traffic also leads to increase in air pollution[28] and other hidden forms of waste.

One hidden type of waste that is often forgotten is the rubber in car tyres. Producing tyres has 'monumental impacts, ranging from continued deforestation to the climate-harming fossil fuels used to make synthetic rubbers to the assembly process . . . [and] what's also becoming increasingly clear is that as the rubber[29] wears, tyres throw off tiny plastic polymers that often end up as pollutants in oceans and waterways.'[30] This cost is higher when everybody drives a different car, rather than sharing a bus or train.

The associated costs of mobility are immense and highlight that there is a need to find sustainable solutions to mitigate waste. An example that policymakers can implement is the SIMSystem (Seamless Integrated Mobility), which 'aims to promote more interoperability between modes of transportation in order to avoid a proliferation of potential uncoordinated or conflicting investments, assets, standards, rules and technologies'.[31] Basically, it aims to help increase overall mobility by making smarter public transport decisions. Other potential approaches, such as electronic buses used in Latin America,[32] are being implemented and can make a significant difference.

Sustainable/Smart Cities

'As population growth increases consumption and waste, managing this waste is becoming an ever greater challenge. The internet of things can be used to develop smarter and more effective ways of managing and reducing waste.'[33]

Sustainable buildings and architecture are being promoted as a key step towards a sustainable future. The negative effects of current unsustainable practices include climate change and the urban heat island effect[34] where the temperature of the city is higher than the areas surrounding it.[35] The resulting effects on people living within urban areas are heightened due to unplanned growth and environmental degradation.[36] These are profound impacts that are felt more and more throughout, in part due to the scale of master plans,[37, 38] along with an increase in population and poorly managed practices within cities.

Notably, twenty-two of the world's top thirty polluted cities are in India.[39] This highlights how a country with over 1.3 billion people[40] can produce profound amounts of waste despite disposable income being lower than many economically developed nations.[41] Generally, 'waste generation varies as a function of affluence'.[42] Yet, notably 'regional and country variations can be significant, as can [waste] generation rates within the same city',[43] highlighting that managing waste and receiving accurate information about the overall system changes how polluted cities may become.

Impacts of Waste That Modern Cities Face

'Municipal solid waste management is one of the major environmental problems of Indian cities. Improper management of municipal solid waste (MSW) causes hazards to inhabitants. Various studies reveal that about 90 per cent of MSW is disposed of unscientifically in open dumps and landfills, creating problems to public health and the environment.'[44]

If we focus on plastic waste alone, it should be noted that:

- Humanity is producing over 300 million tonnes of waste annually,[45]
- Half of this is single-use products,[46]
- Approximately 8 million tonnes of plastic ends up in the ocean each year,[47]
- The take-make-dispose system practised in all of those twenty-two most-polluting cities is contributing (together with all other locations worldwide) to the equivalent of five grocery bags worth of waste on every 30 cm of every country's coastline.[48]

Yet, current research has highlighted that despite the way things are often conveyed (be that via media or other forms of communication) the available facts illustrate that the world is in a better position than presented.[49]

Staying positive can lead to solutions because people within a city are hopeful that what they are doing will make a difference.[50] This is not to say that things cannot be improved, or that they don't need to. They most certainly do,[51] but all the actors within the system must be utilized and valued if a society is to transition from a linear system to a circular one that mirrors the natural world, while utilizing the technological advancements and innovations.

Understanding and appreciating people as a resource will enable those in power to change how a city functions. Elected officials, for example, can do so much with information that illustrates the situation clearly and accurately. There are an estimated 1.5–4 million waste pickers in India[52] who sort, segregate and evaluate the waste collected from the streets of towns and cities. They manage to recycle up to 20 per cent of the entire country's waste[53] and know and understand the situation better than most people. Waste pickers are valuable human resources who have a vast array of information.

Significantly, India has made, substantial progress[54] in moving towards waste-free practices.

Future-Ready Cities and Stakeholders

'Globally, city economies in India, Vietnam and China have the strongest short-term momentum. The pace and scale of change in these markets is extraordinary . . . While these changes present opportunities, many of these cities are facing challenges to their longer-term development prospects, with strains on infrastructure, high levels of inequality, issues around affordability and environmental degradation. Such rapid transformation is often eye-catching. But it is cities that are investing in a sustainable future, and laying the groundwork for ongoing success, that deserve recognition.'[55]

Here are a few examples of zero-waste cities[56] from around the world:

- **Vancouver, Canada**: The city adopted a zero-waste approach in 2006! As of 2018, after years of discussions, the 'Zero-Waste 2040' proposal was approved. The strategies included in the proposal include reducing single-use plastic, focusing on circular economy, incorporating innovation, reuse, repair and re-manufacturing products to eliminate waste, and composting organic waste.

- **Bute, Scotland**: In 2015, the Isle of Bute was selected as the second town to participate in Scotland's 'Zero-Waste Towns' initiative. The pilot project's 'inspire, educate, empower' community action and circular economy approaches are being implemented, reuse and recycling have increased, and household waste to landfills has dramatically reduced. Even unsold loaves of bread are being put to use—they are diverted from the landfill to Bute Brew Co. brewery where they're recycled into craft beer.
- **Cape Town, South Africa**: When FIFA announced that South Africa was to host the 2010 World Cup, a small community in Cape Town set out to clean the city. Led by NGO Thrive Hout Bay, local schools, residents and businesses came together to work towards a 2010 zero-waste mission. In partnership with the local government, they launched a household waste drop-off programme that proved to be an inspiration for other drop-off programmes across the city.
- **Alaminos, Philippines**: In 2009, the city council, along with support from the Global Alliance for Incinerator Alternatives (GAIA), the community and the local government, implemented composting and waste segregation in a city that once was victim to open dumping and burning of waste.

A similar step forward was taken in Indore, in the state of Madhya Pradesh, which was ranked India's cleanest city in 2019.[57] Three years prior to this recognition, the city's municipal council banned garbage dumps and implemented a strict segregation programme that was initially met by resistance.

'When the waste segregation began, 80 per cent of people would not segregate waste,' a garbage truck driver in Indore noted. The individual workers 'would explain to [people] why they had to do this, and plead with them. If people repeatedly gave unsegregated [waste], we would let our supervisors know and fine those households.'[58] In 2019, the streets of Indore were clean thanks to the help of all stakeholders in the city, and a clear, structured message that used available facts and resources. The future will require key stakeholders, ranging from waste pickers and truck drivers like those in Indore to everyday consumers, business leaders and local and national politicians to act together.

Engaging everyone will allow new, waste-free and sustainable solutions to be implemented, such as green roofs, which are more sustainable than other infrastructure.[59]

The dominant practices in construction in the past, such as discarding building debris in landfills and using implosion techniques for demolition, need to be reimagined. These methods in many cities worldwide have led to waste being expelled into the air, on land and into the water, leading to detrimental impacts on human health and the environment.

Most challenges to human health are caused by factors that can be addressed in sustainable environments such as green cities. Addressing sewage pollution, for example, 'could prevent as much as 25 per cent of the world's disease burden. There is also potential for substantial long-term savings (for all stakeholders) by addressing root causes.'[60] Some of these causes include unplanned expansion of urban areas, poorly designed and managed water and energy sources, and inefficient waste systems that are reliant on assistance from individuals in lower socio-economic groups. These areas require partnerships between the numerous stakeholders[61] within a system to develop a process that will make sustainable cities the norm rather than the exception.

Water pollution occurs, for example, when toxic substances enter water bodies such as lakes and rivers. The substances degrade the quality of water as the 'pollutants [can] seep through and reach the groundwater, which might end up in . . . households as contaminated water [used in] daily activities, including drinking'.[62] Waste from a city's sewers and industrial waste discharge are two of the biggest sources of pollutants, but many more sources are to blame as well. In India, a country with fourteen major, fifty-five minor and numerous small rivers, and over 1.3 billion people relying on access to fresh water, continual contamination of fresh water sources is alarming.[63, 64]

Although many cities in India and other countries have not latched on to innovation like the city of Indore, there are a number of other ways that smaller changes are leading to more tangible solutions. To clean the Ganga, for instance, Cleantech Infra has designed a boat that collects floating waste.[65] Meanwhile, other businesses are focusing on addressing the issue before the pollutants enter the waterways by designing products where 'recyclability is already considered during the design phase of a product. All technical and biological processes must be environmentally compatible'.[66]

Product innovations in the city can lead to consumers adopting environment-friendly practices without a dramatic shift in overall habits. These include avocado seeds that are transformed into disposable bioplastic straws and cutlery in Mexico[67, 68] and the root vegetable cassava being made into a biodegradable bag in Indonesia.[69] Paramount in the success of these products will be an added awareness about why they are being used.

For example, solar panels, which have been more widely adopted than green roofs, are a catalyst for change. Due to consumer demand, solar panels have reduced in price, which is helping wean society away from the use of fossil fuels. In 2018, solar energy output increased by 24 per cent globally. The majority of the demand came from countries in Asia, including China, Japan and India, as well as the USA, Australia and Germany.[70, 71]

Transitions to clean energy and a waste-free environment is pivotal to long-term sustainability. These areas are now becoming more affordable and accessible to people within cities wanting to make a difference.

The full implementation of a circular economy within the city will require large-scale change supported by governments. Sustainability can be achieved with the use of new technology and a desire to shift from methods that have been in place for over a century, but it will require committed stakeholders who are willing to transition to long-term solutions even though there may be short-term pain.

In many locations, there needs to be a strong catalyst, often personified by concerned stakeholders and a high guarantee of potential success in the new system so that those in power do not shy away. Yet, there is no denying that a sustainable city, with better mobility and infrastructure, is a valuable model. However, given the risks, the stakeholders may be a little hesitant.

There are valuable and clear facts available for those in power to remodel cities, especially those that currently consume 60 per cent of the world's energy and generate 70 per cent of the greenhouse gases and global waste.[72] Additionally, there is a promising future ahead if circular systems are put in place[73] in order to create a circular city.[74]

A move from a linear to a circular system is not simple, but there are multiple stakeholders who can make a difference and multiple methods that can be implemented to move away from wasteful practices. Exponential growth and linear processes have led to a tipping point; changes must start to occur now. The earlier these new methods are implemented, the less effect there will be on you or any other individual living within the confines of the city. People within the city, active citizens who are involved with action groups and/or businesses, for instance, are in a position to promote change. This, along with the individual changes discussed in earlier chapters, is where you can make a difference.

What Can You See in Your Waste Now?

'The important thing to remember is that change is about slow and steady progress. The way cities are being reimagined can shape how waste is viewed. Historically, cities have been ingrained with ideas from the past, such as we can throw waste away and no negative consequences will occur. It's time to shake off that image, reassess our cities and look towards all the opportunities that tomorrow will bring if we value our resources and use our experience and science to lead us forward.'

LOOK A LITTLE DEEPER
WITH THIS ACTIVITY!

Activity #3 is a four-part process, on separate question sheets, that builds on everything you have learnt throughout this chapter. The importance of this exercise is to understand and become aware of the processes that can help you transition to a sustainable lifestyle.

ASSESS YOUR
CITY WASTE:

What type of environmental impact do your products/services have?
Join the product/service to the problem using an arrow:

Product/Service:	Problem:
Commuting	High initial investment
Waste collection	Traffic congestion
Green infrastructure	Overburdened waste system

Use the ideas from this sample to assess your waste. The product/service
may relate to more than one problem. Fill in the lines below with your
products/services and environmental problems.

_____   _____

_____   _____

_____   _____

_____

When thinking about the environmental impact, it is important to think of the number of areas that it could
affect. Start on a small scale and work your way out. First, think about what it means to the micro-
environment around you, then think larger and larger until you look at it from a global perspective. It will be
beneficial to undertake some research online or in a library.

What system issues prevent change?

Join the product/service to the issue using an arrow:

Product/Service:

Commuting

Waste collection

Green
infrastructure

Issue:

No government
incentives

No public transport

Inefficient
management

Use the ideas from this sample to assess your waste. The
product/service may relate to more than one issue. Fill in the lines below
with your products/services and environmental issues.

_____   _____

_____   _____

_____   _____

_____   _____

_____   _____

A system is anything associated with the product or service that is interconnected with it, for example, the
manufacturing unit that produces the product or the government office that provides the service. To learn
more, conduct research online or in a library.

ASSESS YOUR
CITY WASTE:

What sustainable options are there to replace it?
Join the product/service to the solution using an arrow:

Product/Service:

Commuting

Waste collection

Green
infrastructure

Solution:

Investment in
technology

New public
transport systems

Decentralized
approach

Use the ideas from this sample to assess your waste. The
product/service may relate to more than one solution. Fill in the lines
below with your products/services and solutions.

_____   _____

_____   _____

_____   _____

_____   _____

_____   _____

A sustainable option is a product or service that will last longer and/or produce less waste. Think about
options such as products made from earth-friendly materials, or those that reduce waste through a supply
chain. To learn more, research online or in a library.

ASSESS YOUR CITY WASTE:

How will you start using the sustainable option?
Join the product/service to the action using an arrow:

Product/Service:	Action:
Commuting	Encourage change in the sector
Waste collection	Lobby for public transport
Green infrastructure	Segregate waste

Use the ideas from this sample to assess your waste. The product/service may relate to more than one action. Fill in the lines below with your products/services and actions.

_____   _____

_____   _____

_____   _____

_____   _____

_____   _____

This last step is all up to you. Make your choice, know the benefits and live a zero-waste lifestyle. You are more likely to succeed with support from your friends and family.

ZERO—WASTE LIBRARY

Sharing Economy

A sharing economy is one that opens up opportunities for all, allowing materials and resources to be used in the best way possible.

- **The Need**: A lot of assets (electronics, tools or equipment) that are owned by individuals are used for only a few hours in a month or even less. An effective sharing economy model helps realize underutilized potential.
- **The Solution**: A localized peer-to-peer sharing platform for electronics, tools or technical equipment, which becomes a thriving marketplace for lenders and borrowers with insurance coverage. The insurance cover helps to keep the system secure and maintains integrity. The benefit is that it offers access to a wide range of high-end equipment at an affordable cost, which can also serve as extra revenue from your idle assets. This creates a strong incentive for the renter to look after the equipment and for manufacturers to create equipment that lasts.

Some examples of a sharing economy in action are:

- Tool Library,
- Repair Café,
- Yulu Bike Share,
- BLive Electric Bike,
- Zip Car,
- Amsterdam's Sharing Economy Action Plan.

Research online to learn more about these examples and to find out more. One of the best sources of knowledge in this regard is the Ellen MacArthur Foundation.

Waste Clean-Ups and Environmental Action Groups

A great way to make a difference in a city is to become involved in groups focused on cleaning the environment. There are many of these on social media or with their own websites. Listed below are some of them:

- **The Ugly Indians**: They are a group of volunteers who help clean up India's streets. You can mail them to get involved.[75]
- **Clean Shores Mumbai**: This is a Project of United Way Mumbai that cleans the beaches of the city.[76]
- Environmental Foundation of India: It works across multiple cities in India, collecting waste from lakes, ponds, beaches and other areas. The focus is on wildlife conservation and habitat restoration.[77]
- **Shuddhi**: This is a volunteer group that focuses on numerous aspects, including clean-ups of rivers and beaches. The focus is also on education, the environment, water, wildlife and disaster relief.[78]
- **Aahwahan Foundation**: This is an NGO that consists of coastal management activists who help to remove waste from beaches. The organization uses beach-cleaning machines and other safety equipment to ensure the volunteers, who help them out, are not harmed.[79]
- **Earthlings**: Here, the focus is on a range of activities, including beach cleaning, tree planting and cleaning up sections of cities, such as stained walls and waste piles on the streets.[80]
- **Mahim Beach Clean Up**: This is a UN Environment Programme-facilitated citizen movement that looks to free Mahim beach of waste.[81]

How to Organize Your Own Clean-Up[82]

- Identify a site (try to choose a place that needs a good clean and ensure that volunteers can get to the site safely),
- Visit the location before the clean-up to make sure it is ready (by checking beforehand you can change the site if you need to),
- Appoint someone in charge who can coordinate,
- Gather supplies and safety equipment, such as bags, masks, gloves or ask your volunteers to bring their own,
- Good planning is key. Make sure you know what you will be doing with recyclables and non-recyclable items beforehand. You can coordinate with nearby recycling centres, for instance,
- Depending on the size of your event, you could get event partners on board,
- Ensure that everyone knows what to do for hazardous waste. For instance, you could have a specific container to place dangerous material in,
- Advertise the event online and in your community to gain volunteers,
- On the day, stay organized and visit your site. Stay safe when collecting the waste and placing into the bag and keep up good levels of hygiene so that everyone leaves as healthy as they were when you began,
- Make sure to take pictures to share the good work that your team has done!

How to Create Your Own Online Petition[83]

This is a really useful thing to know if you want to promote change in your city, such as encouraging a shift to sustainable transport options.

- **Know Your Goal**: Identify the problem you want to address, gain an understanding of who your target audience will be, note down exactly what it is that you are trying to change and work out what your desired outcomes should be.
- **Learn About the Change Maker/s**: There will be specific people within the city who can trigger change. This could be people in the government. Learn about them to know the reasons why they do what they do and what could make them change their mind.
- **Stakeholder Engagement**: Understand the needs of the people on your side and on the other. Learn about everyone's demands, opinions and reasons behind being involved. This can be a basic overview. Of course, the more you learn the better position you will be in to put your case forward.
- **Strategies**: The way you approach your movement will be based on the needs of your campaign. You could mobilize people around you, set up meetings with people in charge or implement other tactics that suit your needs. You can research about methods online if you are short on ideas.
- **Active Engagement**: To make your petition effective, you may want to have meetings with public officials, connect with people via email or visit open government forums. Engagement points such as these will be based on your campaign; the important thing to remember is that if you are actively involved, you may come across more opportunities.
- **Create Change and Report**: With online petitions, or offline efforts for that matter, it is important to place everything into perspective. Learn from the challenges, record your wins and your setbacks. This way you can teach others and learn from many more.

Here are some examples of citizen movements:

- Fridays for Future (there are chapters all over the globe),
- Let India Breathe,
- My Mollem (Amche Mollem).

Waste Management Organizations in Indian Cities[84]

To learn more about the types of waste management happening around you, undertake some research into the organizations involved. A select range is detailed below:

- Antony Waste Handling Cell Pvt. Ltd, which works with municipal solid waste,
- Arora Fibres, which recycles plastic bottles into polyester for packaging,
- Eco-Wise, which works with municipal solid waste for residential, commercial and industrial waste,
- Greenobin, which focuses on waste paper management and office recycling,
- Green Power Systems, which is a waste-to-energy business,
- Hanjer Biotech, which creates green products from mixed solid waste,
- Let's Recycle, which works to collect and divert dry waste from landfills,
- Sampurn(e)arth Environment Solutions Pvt. Ltd, which focuses on decentralizing waste solutions for houses, townships, schools and tertiary organizations. They focus on processing biodegradable and non-biodegradable waste,

- Synergy Waste Management, which works with biomedical waste,
- Vermigold Ecotech, which is a specialist in organic waste and uses vermicomposting methods to reduce waste.

< 249 >

CHAPTER 9

TRAVEL

'Poor waste management—ranging from non-existing collection systems to ineffective disposal—causes air pollution, water and soil contamination. Open and unsanitary landfills contribute to contamination of drinking water and can cause infection and transmit diseases. The dispersal of debris pollutes ecosystems and dangerous substances from electronic waste or industrial garbage puts a strain on the health of urban dwellers and the environment.'[1]

The Perfect Life . . .?

You've escaped. The highway is yours. Nothing can stop you. Well, nothing except lunch. You make a quick stop at a large takeaway chain. You start your car again and throw the packaging into the single bin at the entrance to the drive-through area. Most of it ends up in the hole your throw was aimed for. You leave it all behind and soar ahead of the other cars on your way to the airport.

You could have taken the bus, but despite the proximity of your holiday abode it was a better option to fly. Far more comfortable, you think again. You hand your printed tickets in hand and stamps, labels and cable ties on all of your bags.

You convince the check-in counter attendant to let you carry more items than the weight allowance, assuring her that you will not bring it back on the return journey. You are at a resort, you state, you won't be leaving it and none of the extra weight the check-in counter attendant let you have will be coming back on the return flight. You appreciate yourself for twisting the truth so well. You will, of course, leave all of the small soaps and shampoo bottles you got from your last hotel behind and replace it with new bottles from this resort. You smile as the check-in counter attendant lets you through.

Next, you remove and put on most of your clothes and gear at the screening gates. Then you sit in the waiting area until you notice the shops. There are nice trinkets for sale, you notice. You know that you do not need them and that they are in fact similar in design to those from the area you are travelling to, but that does not matter. If you buy them here, you will not need to leave the resort at all.

You know that outside the confines of the five-star location are dusty roads and people who cannot afford to stay there. You really don't want to meet them and bargain with them on a relaxing trip. They should focus on farming or fishing or whatever they do, you think, as you buy the trinkets from the big retailer and stuff them into your full bag that you had, just moments ago, assured the check-in counter attendant you would not add anything to.

You hear an announcement for your flight. You grab your bag and a couple of in-flight essentials from the flight attendants. You wonder for a moment whether you would have got any of this on the bus or a train. You smile. That is no way to travel! You've earned your way to travel in any way you wish. Nothing else matters. A cool drink with a straw and a little umbrella floating next to the perfectly purified ice cubes awaits.

What Is in Your Waste?

'Getting a grip on the mountains of solid waste produced by humanity is central to the (UN Environment) assembly's goal of moving earth "towards a pollution-free planet". After all, poorly contaminated rubbish contaminates our air, water and soil, and represents a colossal waste of the planet's finite resources.'[2]

'I want to conclude these insights with two terms that I have found to be really useful. I discovered these while travelling to locations in India and abroad: responsibility and balance. I think about my niece often when I reflect on these things. She is a little older now, compared to when I first spoke about her. I told you about the waste that I saw at my sister's house when I visited her shortly after the little one was born. I really want to be someone who helps my niece have similar opportunities to what I have had. I want her to be able to see the natural world without waste discards on mountaintops, littered across desert plains or bobbing up and down in the water. She did not cause any of those issues, not yet, but she will most likely be confronted by them, like many young people are today, if things do not change.

Travelling allows us to see different aspects of the world. For a brief period, we are able to move away from the bustling city or region we live in. We often choose to end up in a natural area, or a uniquely different environment. The individualities of these locations pull us in profound ways. I have often rushed to a beach, as many others have, with a smile on my face, and climbed to the top of a peak with huffs and puffs on other occasions for an amazing view, but with a sense of accomplishment like no other. As Indians, we are gaining more and more opportunities to do these memorable things as our country develops and our pockets become fuller with disposable income.'

< 254 >

NOW IT'S YOUR TURN TO ASSESS!

Try out Activity #1 with your current knowledge. View this activity as an initial way to see how much you know. Once you have learnt more as we go forward in this chapter, you can return to it in order to see how much your knowledge has developed.

ASSESS YOUR
TRAVEL WASTE:
ACTIVITY #1

Follow the example sheet shown by choosing a few products/services (for services it could be something like transport) you use to assess.

	Example A	**Example B**	**Example C**
Is your mode of transport the most environment-friendly option available?	✓	✗	✓
Is the waste created due to single-use products?	✗	✓	✗
Is the problem with this situation that consumer behaviour changes when travelling?	✗	✗	✓
Do the issues have wider impacts, i.e. on a global scale?	✓	✓	✗

- If yes: share the solutions with friends and family.
- If no: research options.

Use the ideas from this sample to assess your waste. You can draw your own sheet based on this and create other questions for this assessment based on your needs.

What Resources Are Available?

'When we think about resources, it is this "disposability" that I am concerned about in relation to my niece's generation. During my trips to north Karnataka with SELCO Foundation, I often witnessed, while sitting on overnight buses, how mindsets change on the road. We tend to want to constantly keep our homes and immediate surroundings clean, but once we travel outside the confines of our personal boundaries there is a higher disregard for cleanliness or hygiene. Empty bottles and wrappers are left on the sleeper bunks and underneath the semi-reclined chairs. Cigarette butts and food containers are carelessly discarded at the rest stops along the way. The people leaving them knew that they were not coming back. Perhaps they trusted the waste system in place, not knowing that the linear economy has led to waste spreading from the top of Mount Everest to the bottom of the Pacific Ocean. Or perhaps they were oblivious and accustomed to the situation.

Either way, when I visited the villages, I saw the effects. They were similar to what you see in a city but with far more destructive spread. The waste system is not great in my hometown of Bengaluru, but at least there are people to manage it. In smaller towns and communities, managing waste often means burning it or placing it in a drainage ditch, because there is nowhere else for it to go. For the two years that I worked in that region, I witnessed this situation in many locations. I heard stories of how things were in the past and how they had changed with migration and people travelling through.

I found that listening to them allowed me to be open to new cultures, introducing me to new forms of food, dance and handmade textiles. All of these areas were amazing to learn about. Many of these practices were sustainable in the original form. Yet, with the global effects of single-use products changing the way people live, some of their practices had become more damaging to the environment.

What upset me the most was witnessing the farmers spend their disposable income, purchasing Rs 2 snacks, shampoo and detergent sachets that pollute the very soil and water in which they, so diligently and with immense hard work, grow the food that sustains us. There is an opportunity to learn from this microcosm in order to develop new techniques that utilize traditional sustainable practices with modern complexities and innovation.'

TRY THIS ACTIVITY OUT!

In this chapter, there are a number of suggestions that you can use for Activity #2. You can read ahead and return with new knowledge, or assess yourself now before learning a little more and return to this exercise later.

ASSESS YOUR
TRAVEL RESOURCES:
ACTIVITY #2

Learn how you can pack a travel bag for a zero-waste trip and/or pack a bag and go for a zero-waste holiday. Use the recipes in the guide book or research to find more. Fill in each section of this activity to record your achievements.

The resources that I have learnt about are:

 E-tickets for travel on transport.

 I packed a zero-waste bag.

 I travelled for a week, on a zero-waste trip, staying at an eco hotel and using sustainable transport.

The resources needed for this activity are:

 I used my phone for the e-ticket.

 I had all the zero-waste items I needed from lessons in earlier chapters of the guide book.

I donated to an organization to offset my carbon footprint for travel.

Draw a picture from your activity to show friends and family:

Record where you learnt about these resources:

 The guide book's 'Tips and Tricks' section, other chapters in the book and online research.

Record who you shared your success with:

 My partner! We went on the zero-waste trip together. It was amazing, we hiked to the top of mountains and stayed at an eco hotel.

Use the above ideas to assess your resources and then create your own activity sheet.

Moving Towards a More Sustainable Lifestyle

'There are so many things to learn about reducing waste that you are exposed to while travelling. When I spent time with the women of Harobelavadi village, who were making vermicelli noodles while telling me how they were way better than the store-bought one, I had another of my aha moments. Those lovely women are the humans we must support when we travel. Buying unique products from them and being welcomed into their lives offers so many benefits!

One of my favourite learnings while travelling was how to make a traditional roti that I now make at home, in an environment far removed from Harobelavadi village.'

ZERO—WASTE RECIPES

This is the recipe for Holige/Obbattu/Meethi Roti from north Karnataka.[3, 4]

You will need:

- 1 cup maida/all-purpose flour,
- ¼ cup cooking oil,
- ⅛ teaspoon turmeric powder,
- A pinch of salt,
- Water (for the dough, judge it yourself, anything up to ½ a cup is enough).

< 260 >

For the stuffing, you will need:

- ½ cup toor dal or chana dal,
- ¾ to 1 cup grated jaggery,
- ¼ cup grated coconut,
- 2 cardamoms,
- Water (as required).

Follow these directions for a truly wonderful meal:

For the Dough

- Add the flour, turmeric and salt into a bowl. Mix well and create a small dent in the centre to add a little water until the dough is sticky,
- Add a teaspoon of oil to the dough and knead it with your palm for about a minute. Add more oil whenever the dough starts sticking to your hand. Keep repeating this until the dough is soft. This will take anywhere from 5–10 minutes,
- Place the dough back into a bowl. Pour any remaining oil over the dough to allow it to soak the oil. Close the bowl with a lid (or place a plate over the top to seal it) and leave for 6–8 hours,

For the Stuffing

- Boil the dal in a bowl with 4 cups of water. Ensure the dal blossoms but does not become mushy. Do this in an open pot and not a pressure cooker to gain this consistency,
- If there is any water left in the dal, drain it into another bowl to make obbattu saaru,
- Grind the cooked dal, grated coconut, grated jaggery and cardamom into a paste using a mixer,

- Take the stuffing and roll them into lemon-sized balls. Cover them and set aside,
- Combining the dough and stuffing. Once the dough has rested, take a handful of dough (half the size of one of the stuffing balls). Place the ball on a greased obbattu sheet and spread it lightly so that the corners are thinner than the centre,
- Place the stuffing ball in the middle and cover it with the dough,
- Press it down, distributing the stuffing evenly,

For Cooking

- Pour some oil on to a dosa tawa and wait for it to get hot. Once that is done, take the obbattu sheet and invert it over the tawa until the holige drops on to the heated tawa. Cook both sides on a medium flame until it is golden brown,
- Once it is cooked you can add milk and ghee to add to the taste.

'For the Harobelavadi community, like you, me, and even my little niece, travel has opened doors for new experiences. It has allowed for intermixing in very beneficial ways that were not previously possible. We have an opportunity to learn from one another and gain experiences.

Yet, if we continue to see travel as a reward where we can do anything that we want with no consequences, where does that leave our future generations?

We are responsible for our choices on the road. We can view the ill-effects of bad choices at bus stops or in villages, on top of mountains or in rivers and streams. Simple steps forward can involve changing our practices at hotels by not using small packets of disposable items, or while on the road by thinking and caring about what we do. Broader still, understanding the repercussions of our actions will allow us to treat the environment that we're in with respect. If we leave the places we visit in a better state than when we arrived, that is taking responsibility.'

ZERO—WASTE TIPS AND TRICKS

- Commute sustainably, e.g., take trains, buses, boats, etc.,
- Say no to the small single-use packets at hotels (such as freebies that often come in little plastic containers),
- Shop local while travelling, such as from artisans and markets, and avoid items wrapped in plastic,
- Carry reusable products like food containers, water bottles, cutlery, straws and personal-care items,
- Carry your own napkin and unpackaged snacks,
- Use PDF versions of tickets instead of printouts,
- Carry your own headsets to avoid single-use items wrapped in plastic, which need additional resources to clean and package it for next use,
- Pre-download your books or borrow from the library instead of buying new ones (you can do the same for podcasts, too),
- Opt for reusable luggage tags, or make your own version at home ahead of time,
- Shop local while travelling (local farmer's markets for some strawberries, anyone?),
- Download Bea Johnson's 'Bulk Locator App'[5] (it shows stores that offer bulk products within your vicinity, which is really useful while travelling to new areas!).

'I want to share a quick travel story from when I was seventeen. My elder sister took me scuba diving to this really beautiful island, Netrani Island, in the Arabian Sea. A travelling experience like that really puts things into perspective, making you realize how small, almost insignificant, you are in comparison to the vast and beautiful planet that supports our lives. I may not be religious, but you can say that I felt a spiritual connection to nature.

In a way, this isn't hard to believe because as humans we are meant to coexist with nature and live within its folds rather than exploit it. Nature became my teacher, therapist, calm and solace. A walk in the park or a stroll on the beach has always been my go-to solution when life gets too tough (I'm sure you can all relate to this!). The positive effects of surrounding ourselves with nature has scientific evidence to back it.[6]

I have also found that travelling with like-minded individuals or groups can have quite a great influence. For example, a community that I really want to hike with, India Hikes, constantly encourages hikers to pick up waste along the way, adding to the testament of taking the responsibility and cleaning up rather than approaching it with the mindset of it being someone else's waste or someone else's problem.'

< 264 >

TRAVEL - LINEAR GLOBAL IMPACT

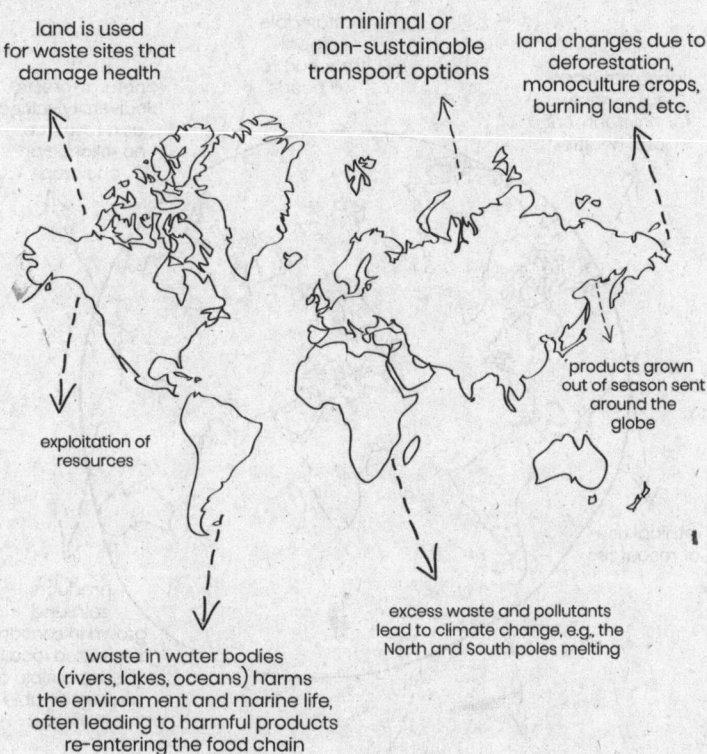

land is used
for waste sites that
damage health

minimal or
non-sustainable
transport options

land changes due to
deforestation,
monoculture crops,
burning land, etc.

exploitation of
resources

products grown
out of season sent
around the
globe

waste in water bodies
(rivers, lakes, oceans) harms
the environment and marine life,
often leading to harmful products
re-entering the food chain

excess waste and pollutants
lead to climate change, e.g., the
North and South poles melting

TRAVEL - CIRCULAR GLOBAL SOLUTIONS

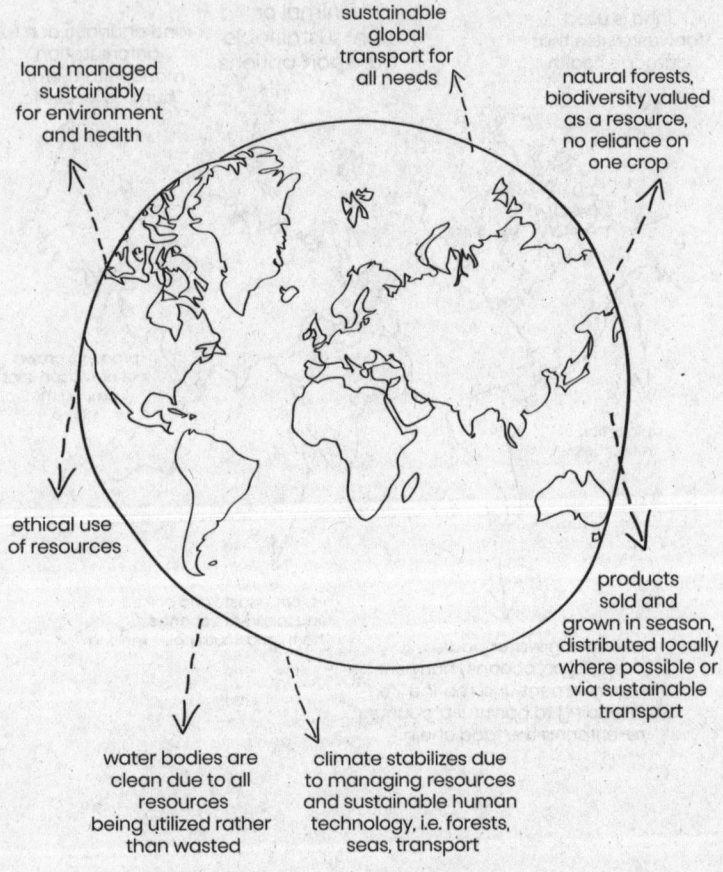

sustainable global transport for all needs

land managed sustainably for environment and health

natural forests, biodiversity valued as a resource, no reliance on one crop

ethical use of resources

products sold and grown in season, distributed locally where possible or via sustainable transport

water bodies are clean due to all resources being utilized rather than wasted

climate stabilizes due to managing resources and sustainable human technology, i.e. forests, seas, transport

Why Is It Harmful?

> 'We've had an incredible journey to get here. By now you know how much our individual choices at home, or in the community, really affect the whole world. It's pretty amazing, isn't it?
>
> Learning about the negative impacts of waste and the fact that there are organizations and individuals trying to make a difference in order to reduce the impact of unsustainable habits is vital. It allows you to be an active citizen and to share your knowledge and learn from others. There are always things that you can do to make a difference. Enjoy this last parcel of information. I will see you in our last section with one final guiding thought.'

The question of whether travel is good for the planet or not is a challenging one to answer. Certainly, some modern ways of travel do emit large amounts of waste into the environment. Jet fuel would not have been produced had people chosen not to travel such distances at such speeds. However, without travel people would not be connected the way they are today,[7, 8] which results in the encouragement to learn about new cultures and ideas while sharing thoughts on sustainable practices. Knowledge sharing and understanding what others are doing works for governments, businesses and individuals. This means that partnerships can be formed to improve upon innovations or practices that businesses are pursuing. If travelling is avoided completely, the interaction of communities, the sharing of knowledge and gaining of empathy about other people's situations, which would enable you to have informed opinions that can lead to positive outcomes, is far less likely to occur.

As discussed in the previous chapter, there are many complexities when thinking about travel and the world as a whole. Therefore, the small-scale effects of the waste crisis are important to think about first. These are things you have direct control over, like buying high-quality items that will last to avoiding single-use products in order to limit your impact. Additionally, gaining an awareness of the larger tangible impacts and what businesses, communities and other organizations around the world are doing to counter this situation is vital to live a more sustainable life. Your individual choices can assist in a transition towards a waste-free planet for all. Notably, for a transition to more earth-friendly practices, alternate options need to be available. Some examples of this type of change are listed below.

Negative Global Impacts from Mismanaged Waste

'Where waste cannot be avoided, recovery of materials and energy from waste, as well as remanufacturing and recycling waste into usable products, should be the second option. Recycling leads to substantial resource savings.'[9]

One specific area to consider is waste from materials that consumers wear or use while travelling. This includes backpacks, suitcases and clothing. These items cause more greenhouse gas emissions than air travel with the plastic in clothing causing more damage to the oceans than cosmetics.[10] In part, this is due to the trend of fashion as discussed in Chapter 2. The 'rates at which clothing and other textiles are purchased are high, as are the associated rates at which end-of-use items arise. Ensuring that methods and systems are in place to permit and encourage items deemed to be end-of-use by one person to be utilised to their full potential . . . is desirable'.[11]

On a similar note, plastic waste,[12] some of which comes from materials such as clothes and single-use products that do not decompose, is also having adverse effects on the planet and waterways. The 'Great Pacific Garbage Patch is a collection of marine debris in the North Pacific Ocean[13] . . . [it] spans waters from the west coast of North America to Japan'[14] and was first discovered in 1997.[15] The area creates a vortex that spins with a still central area, which drags waste to the inside of the patch.[16] The 'spinning debris are linked together by the North Pacific Subtropical Convergence Zone . . . where warm water from the South Pacific meets up with cooler water from the Arctic. The zone acts like a highway that moves debris from one patch to the other.'[17] There are no clear figures on how much waste is in this patch, partly because not all of it floats on the surface.[18, 19, 20]

The effect of wasteful practices is witnessed throughout the marine environment[21, 22] and also on land in less economically developed countries. In those locations, between 400,000 and 1 million people die each year because of mismanaged waste that washes up in their local environments partly due to increasing use of single-use material. This waste comes directly from businesses and people not managing the products[23] well after consumers no longer deem it necessary. Products are simply disposed of using the linear economy's process, instead of considering a circular economy that enables these situations to be fixed.

Positive Impacts from Sustainability Focused Organizations

'Stakeholders and governments should know that SWM (solid waste management) is a complex system that involves environmental, social and economic issues, which should be evaluated holistically for improving the life cycle of waste, reducing water, soil and air contamination due to open burning and open dumping, practices widespread worldwide.'[24]

A notable move towards sustainability in terms of travelling gear, e.g., backpacks and jackets, has been implemented by

Patagonia, a company that emphasizes that keeping products that consumers own is 'the single best thing [humanity] can do for the planet' through their 'Worn–Wear' philosophy.[25] They are focused on this circular environmental action and even have a repair truck that travels to locations to fix any worn-out gear so that it can be utilized for longer, or repurposed. This simple act is a key step forward. There are many other businesses pursuing circular practices, as represented in the Notes section,[26] to limit the amount of waste entering into the environment. The Ocean Cleanup, a Dutch NGO, has developed 'technologies to extract plastic pollution from the ocean',[27] addressing the symptoms of wasteful practices particularly in the Great Pacific Garbage Patch but also in rivers and other water bodies in Asia, as of December 2019.[28]

Impacts of an Unsustainable System

'Essentially, a landfill is a large, full trash bag that just sits in your kitchen. As more bags accumulate, your kitchen will become full and inhospitable . . . In short, it is not a sustainable form of waste management, as future generations will have to address the landfill problem that we are creating today.'[29]

Moving the focus away from physical waste, its impact on people due to wasteful practices within a linear economy can result in an extreme threat, such as climate change.[30] This can be due to many issues like burning clothing, wasting food and commuting in an unsustainable fashion when other methods are available.[31]

Air pollution has led to warmer weather and a reduction of rainfall around the world. Drought-like conditions in certain districts in India, where people are struggling, have led to dire consequences, such as high rates of farmer suicide in the state of Maharashtra.[32, 33]

Notably, the global impacts of waste are often felt outside of everyone's usual sphere of consciousness, especially when living in a city. Even though you may not feel connected to people in far-off regions, your actions contribute in various ways to their lives. The interconnected nature is evident in threats such as fires, floods and droughts, which are caused by decisions made in other parts of the world by people and businesses that are functioning within systems that are not accountable to the planet.

Active Citizen Involvement

'The highest priority, avoiding and reducing the generation of waste, encourages the community, industry and government to reduce the amount of virgin materials extracted and used. The goal is to maximize efficiency and avoid unnecessary consumption.'[34]

Consumers worldwide are demanding sustainable travel options, which places pressure on traditional travel businesses. It is establishing a growing industry, that of sustainable travel, which will help to create jobs and lift people out of poverty. Commonly, these eco-conscious options turn out cheaper in the long run and benefit local communities. Many communities, in fact, are demanding it. For example, Copenhagen's 'localhood strategy', through its 'End of Tourism As We Know It' campaign,[35] focuses on enhancing travellers' and locals' experiences by encouraging more connections between visitors and residents. Similarly, employees of travel businesses want sustainable options for themselves and their clients, too. They, just like the people travelling, are becoming more aware of the overall situation and the benefits of earth-friendly practices.[36]

Large-scale changes can be promoted through businesses and organizations that pursue sustainable practices and people like you becoming active and conscious travellers.

A model that is useful to think about is that of 'stakeholder capitalism', as proposed in the Davos Manifesto by the World Economic Forum,[37] which may enable more sustainable practices globally:

> Generally speaking, [there are] three models to choose from. The first is 'shareholder capitalism', embraced by most Western corporations, which holds that a corporation's primary goal should be to maximise its profits. The second model is 'state capitalism', which entrusts the government with setting the direction of the economy, and has risen to prominence in many emerging markets, not least China. [While] 'stakeholder capitalism' . . . positions private corporations as trustees of society, and is clearly the best response to today's social and environmental challenges . . . But to uphold the principles of stakeholder capitalism, companies will need new metrics. For starters, a new measure of 'shared value creation' should include 'environmental, social and governance' (ESG) goals as a complement to standard financial metrics . . . The second metric that needs to be adjusted is executive remuneration . . . In the new stakeholder paradigm, salaries should instead align with the new measure of long-term shared value creation. Finally, large companies should understand that they themselves are major stakeholders in our common future . . . (the businesses should) work with other stakeholders to improve the state of the world in which they are operating. In fact, this latter proviso should be their ultimate purpose.[38]

Clearly, this suggestion is still in the early stages. It will need some time before businesses in the transport,[39] clothing, food packaging and accessories sectors, among others, are able to develop new metrics that mirror the core ethos of the suggestions. For the moment, however, it is worth questioning, where does this model leave your travelling experience?

Knowledge Sharing For a Sustainable Future

'Common projects should be introduced at a global level in order to reduce the environmental contamination and health issues due to waste open dumping and burning. Authorities and the actors involved in waste management should be aware of the global issues which are affecting sustainable development, providing such information to the population for spreading awareness and its inclusion in recycling and prevention activities.'[40]

A large part of creating a world that implements the circular economy will be personal, formed on the basis of decisions based on an understanding of the effects of individual and wider-scale actions.[41] It is also about knowing that you can make a difference with small changes. Acknowledging that all of these factors are nuanced is important, too, as is learning about changes to the waste system that are occurring globally.

For example, people in the Philippines are now able to swap collected plastic waste for bags of rice. This limits the amount of waste entering the environment and allows the poorer communities better access to more food[42] as an interim solution to a major global issue. More sound solutions take place as mitigation rather than adaption. The Mumbai airport has gone plastic-free,[43, 44] while the San Francisco airport has banned plastic bottles.[45] Elsewhere in the world, experiments have been undertaken in Samoa to see if 'workable, economically viable, socially empowering and sustainable scenarios for repurposing and upcycling plastic waste' can be found.[46] Many of these areas have seen a positive change because of knowledge sharing and partnerships.

Several of the most dramatic events that raised awareness about wasteful practices were witnessed in 2019, including fires in the Brazilian and Bolivian Amazon,[47] and also throughout

Australia.[48] Additionally, climate change activists who promote renewable forms of energy[49] and heightened coverage about the waste in the ocean also played a part.[50]

As discussed in the earlier chapters, being able to create a knowledge base about best practices for the environment and for people is paramount. It is a method that the circular economy promotes because of the interconnected nature of the model.

Similarly, positive steps can be profoundly effective if a sustainable system is properly implemented and people around the world have access to information about it. A key example of how positive change and awareness can address global issues is, noted in the World Economic Forum passage above is the Greta Thunberg effect. 'The young Swedish climate activist has reminded [people] that adherence to the current economic system represents a betrayal of future generations, owing to its environmental unsustainability. Another [related] reason is that millennials and Generation Z no longer want to work for, invest in, or buy from companies that lack values beyond maximizing shareholder value. And, finally, executives and investors have started to recognize that their own long-term success is closely linked to that of their customers, employees and suppliers.'[51] All of these areas have seen positive changes towards a circular economy after 2010.[52] This will lead to better ways to travel and interact with the world, both on a global and personal level.

In closing, I'd like to reiterate that many of the ways to reduce waste have developed from knowledge sharing, which enables people to manage resources better. Knowledge itself is a resource, just like the planet, products and people are. It needs to be used by all stakeholders to find solutions to the global waste crisis. It will enable humanity to find a balance with nature, and hopefully help to reverse the damage caused.

While you are on the road, utilize your own knowledge and lead by example with your choices. You now know how and why the impacts of your decisions and the amount of waste that is produced throughout the world are linked.

Each part of the planet and every single person is vital to an improved system, no matter if they are students, waste pickers, business managers, truck drivers, entrepreneurs or farmers. What matters is you!

What Can You See in Your Waste Now?

'I want to leave you with a concluding thought about the big and little things of the zero-waste journey.

Travel can teach us about balance in the world. Whether it is the way people in villages value and cherish the soil, or recognizing what being open to new ideas can mean. It can even help to highlight our biases. If we're open to it, we can change our mindsets through these lessons. Gaining new perspectives can show us that we can afford to spend a weekend on a beach or sit on a patch of grass reading a book as the sun sets below a hilltop in the countryside rather than being focused on academic or professional pursuits alone. It can help us take the blinkers off from our fast-paced world. Additionally, the beauty in the journey to the "other" is that it can also show us what is out of balance.

There are many locations in India and the world that are feeling the impact of a linear waste system. By visiting these locations, we can become more aware. We can even volunteer to help. We can speak, we can listen. We can share in their hardships, and we can find solutions. There may be things that are out of our hands, such as deforestation, but there are small choices that are within our grasp, such as using a bamboo toothbrush or choosing products without palm oil as an ingredient.

I have learnt so many things by listening to the people I have had the good fortune to meet on the road, as you probably have, too. Firstly, to move towards a more sustainable environment we need to be aware of our actions on our planet. Secondly, we must be mindful of our choices and the fact that having more disposable income does not mean that we should move to a more disposable (wasteful, irresponsible, negligent) mindset when travelling.

If we do, we are causing damage that my niece or your nephew, son or daughter will see. They will not necessarily be able to view green pastures and bright blue water the way we do. Certainly not if we continue to choose to allow waste into the environment. We owe it to them to try and balance our desire to travel and see the world with a mindfulness that helps us choose adventures sustainably. We have a responsibility to make small, tangible choices that may not be a catalyst for worldwide change, but it can be a good start. It may not drastically reduce the amount of waste on a beach, but it can make a dent. In a country with over a billion people, there are enough of us to make a monumental difference.

Such a change will require time, patience, perspective, awareness and good planning. They are a good starting point to move towards being a responsible traveller who can encourage people and the wider, interconnected global society, whom we encounter on the road, to move from a disposable method to an environmentally friendly, circular approach. There is still plenty to learn about how to transition to a global circular economy, but we have more than enough information to start. If my niece, who I love so much, is anyone to go by, there is certainly reason enough to try.'

< 276 >

LOOK A LITTLE DEEPER
WITH THIS ACTIVITY!

Activity #3 is a four-part process, on separate question sheets, that builds on everything you have learnt throughout the chapter. The importance of this exercise is to understand and become aware of the processes that can help you transition to a sustainable lifestyle.

< 277 >

ASSESS YOUR
TRAVEL WASTE:

What type of environmental impact do your products/services have?
Join the product/service to the problem using an arrow:

Product/Service: **Problem:**

Transport   Mass-produced
                                                                       products

Accommodation   Carbon emissions

Shopping   Unsustainable
                                                                       practices

Use the ideas from this sample to assess your waste. The product/service
may relate to more than one problem. Fill in the lines below with your
products/services and environmental problems.

_____	_____
_____	_____
_____	_____
_____	_____
_____	_____

When thinking about the environmental impact, it is important to think of the number of areas that it could
affect. Start on a small scale and work your way out. First, think about what it means to the micro-
environment around you, then think larger and larger until you look at it from a global perspective. It will be
beneficial to undertake some research online or in a library.

ASSESS YOUR
TRAVEL WASTE:

What system issues prevent change?
Join the product/service to the issue using an arrow:

Product/Service:

Issue:

Transport

Waste on land and ocean

Accommodation

Climate change

Shopping

Environmental footprint

Use the ideas from this sample to assess your waste. The product/service may relate to more than one issue. Fill in the lines below with your products/services and environmental issues.

_____   _____

_____   _____

_____   _____

_____   _____

_____   _____

A system is anything associated with the product or service that is interconnected with it, for example, the manufacturing unit that produces the product or the government office that provides the service. To learn more, conduct research online or in a library.

ASSESS YOUR
TRAVEL WASTE:

What sustainable options are there to replace it?
Join the product/service to the solution using an arrow:

Product/Service:

Transport

Accommodation

Shopping

Solution:

Support local economy

Offset carbon

Eco hotels and/or resorts

Use the ideas from this sample to assess your waste. The product/service may relate to more than one solution. Fill in the lines below with your products/services and solutions.

_____   _____

_____   _____

_____   _____

_____   _____

_____   _____

A sustainable option is a product or service that will last longer and/or produce less waste. Think about options such as products made from earth-friendly materials, or those that reduce waste through a supply chain. To learn more, research online or in a library.

ASSESS YOUR
TRAVEL WASTE:

How will you start using the sustainable option?
Join the product/service to the action using an arrow:

Product/Service:	Action:
Transport	Buy from small businesses
Accommodation	Offset carbon emissions by tree planting
Shopping	Stay in earth-friendly locations

Use the ideas from this sample to assess your waste. The product/service may relate to more than one action. Fill in the lines below with your products/services and actions.

_____   _____

_____   _____

_____   _____

_____   _____

This last step is all up to you. Make your choice, know the benefits and live a zero-waste lifestyle. You are more likely to succeed with support from your friends and family.

ZERO—WASTE LIBRARY

The 10 x 10 Challenge: A Minimalism Activity[53]

This is an activity that you can incorporate into a zero-waste trip, or your life in general. Have fun and try out the challenge! For sustainable travel bags and trips, try and incorporate the lessons you learnt earlier in this guidebook.

- Choose ten pieces of clothing from your closet,
- Style the items in ten different looks (you could make a cool game by mixing-and-matching your clothes each day),
- Undertake the challenge across ten days (not necessarily consecutive). The challenge is to utilize your resources in new and creative ways.

Once you have completed this challenge with your clothes, you can implement this in other areas of your life, such as how you carry food on your trip, alternating types of transport on certain days of a holiday and much more. Success is in your creativity!

Indian Adventures with a Social and Environmental Ethos

- **India Hikes**: This is an organization of like-minded nature enthusiasts, hikers and adventurers who aim to leave the mountains in a better state than they found it. They take compostable bags with them to collect any waste that they find. It's an amazing community that grows every year.[54]
- **Ride to Lite**: This is a marathon cycling tour across Arunachal Pradesh. It is India's first full-fledged cycle tour, which is an initiative to provide basic lighting to north-east India's most far-flung inhabitants.[55]

- **Go Heritage Runs**: It organizes fun runs across heritage locations in India. They are a planned series of runs that include culturally important locations in scenic areas.[56]
- **Green Venture**: This is an experiential and educational start-up that brings together knowledge, wisdom, learning in biodiversity, horticulture, gardening, trees, insect-plant relationships, etc., connecting and engaging individuals to the natural world.[57]

Organizations with a Global Reach

There have been a number of organizations mentioned throughout the guidebook, here is a list of several more that are worth learning about. These organizations are focused primarily on waste. There are many more that address waste reduction. Do some research online to learn more. Volunteering to gain experience and/or simply help to achieve the ambitions of these enterprises is a worthwhile travel experience.

- **Plastics for Change**: This was launched with a mission to use plastic waste as a resource for addressing social issues . . . With over 2 billion people living on less than US $2 a day, there is an enormous opportunity to reduce poverty through recycling. [They] have been on a mission ever since—to bring recycling infrastructure to developing regions and creating jobs for some of the most marginalized members of society.[58]
- **rePurpose**: This is a movement of conscious consumers and businesses going plastic-neutral by financing the removal of ocean-bound plastic worldwide. [They] are here to reinvent the wheel of the world's resource economy— one where our duty to protect the planet is ethically shared among manufacturers, consumers and recyclers.[59]
- **Plastic Pollution Coalition**: This is a growing global alliance of individuals, organizations, businesses and policymakers

working towards a world free of plastic pollution and its toxic impacts on humans, animals, waterways, oceans and the environment.[60]

- **Plastic Soup Foundation**: This aims to make everyone familiar with the phenomenon of plastic soup and to stop it at its source. They are a single-issue organization focused entirely on plastics. With a small, committed and passionate team of about twenty people, they give everything to achieve the goal: no plastic in our water or our bodies! They are based in Amsterdam, but their mission has not only been embraced in the Netherlands but also in the USA, the UK and India.[61]

- **GA Circular**: This is a research and strategy firm specializing in waste management and recycling. They aim to 'create a world without waste by driving the transition towards a circular economy in Asia'. They do this by enabling industries, investors, government and multilateral agencies and global foundations to unlock business opportunities from fast growing streams of food and packaging waste.[62]

FURTHER READING:
RECOMMENDED RESOURCES

Nature Learning

Reading and learning is a fantastic pastime. Listed below are several books that speak about the value of nature and understanding the way the world works.

- *Soil Not Oil* by Vandana Shiva: The book explains that a world beyond dependence on fossil fuels and globalization is both possible and necessary. Condemning industrial agriculture as a recipe for ecological and economic disaster, the author champions the small, independent farm: their greater productivity, their greater potential for social justice as they put more resources into the hands of the poor, and the biodiversity that is inherent to the traditional farming practiced in small-scale agriculture.[1]

- *Environmentalism: A Global History* by Ramachandra Guha: The author, an acclaimed historian of the environment, draws on many years of research in three continents.

He details the major trends, ideas, campaigns and thinkers within the environmental movement worldwide.[2]

- *The Climate Solution* by Mridula Ramesh: Drawing on her extensive practical and investing experience, she explores myriad facets of a raging issue: why women are peculiarly affected by a warming climate; how climate change poses a security threat to the Indian state; why just focusing on green sources of power is an incomplete solution for India; how managing waste can create lakhs of urban jobs; and how households can cope in a 'Day Zero' water situation.[3]

- *The Unquiet Woods: Ecological Change and Peasant Resistance in the Himalaya* by Ramachandra Guha: This is a path-breaking study of peasant movements against commercial forestry, bringing the story of Himalayan social protest up-to-date, reflecting the Chipko movement's continuing influence in the wider world.[4]

- *Silent Spring* by Rachel Carson: This book was first published in three serialized excerpts in the *New Yorker* in June 1962. The book appeared in September of that year and the outcry that followed its publication forced the banning of DDT and spurred revolutionary changes in the laws affecting our air, land and water. Carson's passionate concern for the future of our planet reverberated powerfully throughout the world, and her eloquent book was instrumental in launching the environmental movement. It is without question one of the landmark books of the twentieth century.[5]

- *Cradle to Cradle: Remaking the Way We Make Things* by Michael Braungart and William McDonough: 'Reduce, reuse, recycle', urge environmentalists; in other words, do more with less in order to minimize damage. But as architect William McDonough and chemist Michael Braungart point out in this provocative, visionary book, such an approach only perpetuates the one-way, 'cradle to grave' manufacturing model, dating back to the Industrial Revolution, that creates such fantastic amounts of waste and pollution in the first place. The book questions the belief that human industry must damage the natural world.[6]

- *The Last Child in the Woods* by Richard Louv: In this influential work about the staggering divide between children and the outdoors, child advocacy expert Richard Louv directly links the lack of nature in the lives of today's wired generation—he calls it nature-deficit—to some of the most disturbing childhood trends, such as the rises in obesity, attention disorders and depression.[7]

- *Deep Economy* by Bill McKibben: In this book, McKibben offers the biggest challenge in a generation to the prevailing view of our economy. For the first time in human history, he observes, 'more' is no longer synonymous with 'better'— indeed, for many of us, they have become almost opposites. McKibben puts forward a new way to think about the things we buy, the food we eat, the energy we use, and the money that pays for it all.[8]

- *This Changes Everything* by Naomi Klein: The book asks you to forget everything you think you know about global warming. The really inconvenient truth is that it's not about carbon—it's about capitalism. The convenient truth is that we can seize this existential crisis to transform our failed economic system and build something radically better. Klein, the author of global bestsellers, tackles the most profound threat humanity has ever faced: the war our economic model is waging against life on earth.[9]

- *No One Is Too Small to Make a Difference* by Greta Thunberg: This book brings you Greta in her own words. Collecting her speeches that have made history across Europe, from the UN to mass street protests, *No One Is Too Small to Make a Difference* is a rallying cry for why we must all wake up and fight to protect the living planet, no matter how powerless we feel.[10]

- *Hot, Flat & Crowded* by Thomas Friedman: In this brilliant, essential book, Pulitzer Prize-winning author Thomas L.

Friedman speaks to America's urgent need for national renewal and explains how a green revolution can bring about both a sustainable environment and a sustainable America. Friedman explains how global warming, rapidly growing populations, and the expansion of the world's middle class through globalization have produced a dangerously unstable planet.[11]

• *Flammable* by Javier Auyero and Debora Alejandra Swistun: The books is based on archival research and two and a half years of collaborative ethnographic fieldwork. It examines the lived experiences of environmental suffering. Despite clear evidence to the contrary, residents allow themselves to doubt or even deny the hard facts of industrial pollution.[12]

• *Environmental Inequality* by Andrew Hurley: By examining environmental change through the lens of conflicting social agendas, Hurley uncovers the historical roots of environmental inequality in contemporary urban America. Hurley's study focuses on the steel mill community of Gary, Indiana, a city that was sacrificed, like a thousand other American places, to industrial priorities in the decades following World War II.

More to Learn About Waste in India

Listed below are four more resources that have been produced in India. You can use them to continue learning more about waste and to share with your friends and family.

• **Trashonomics**: This is a simple guide to solid waste management that can be included as a supplement to the environmental science subject for middle-school students and above.[13]

• **Upcyclers Lab Boardgames**: Started in 2018, Upcycler's Lab uses behavioural economics and play-based learning to change mindsets around the environment. This is done using board games and puzzles based on concepts such as waste segregation and ocean conservation. Most of their products cater to children between two and nine years of age.[14]

- *Hidden Kingdom* by Nirupa Rao: This is a book on the fantastical plants of the Western Ghats—with hand-drawn illustrations set to rhyme—intended for children and adults alike.[15]

- *Let's Talk Trash: Activity Book* by Shubhashree: 'A little illustrated handbook by Shubhashree Sangameswaran with simple, everyday ideas towards a less messy world. If you're looking to go zero waste, this book is a starting point. Also, a brief look at the past to see what lessons we can learn from our earlier generations and how they were pretty effortlessly #zerowaste, even before it became a hashtag.'[16]

International Organizations That Publish Informative Articles about Nature and Waste

The organizations mentioned here are well known and have global reach and impact. Following their research, articles, successes and learnings will enable you to understand what is happening around you and why it matters for you.

- National Geographic,
- World Wildlife Foundation,
- World Health Organization,
- United Nations (including all of their organizations, like the United Nations Environment Programme),
- The Nature Conservancy,
- World Research Institute,
- World Economic Forum.

ACKNOWLEDGEMENTS

I would like to thank my mum, Afshan Mansoor, and my sisters Sabah and Iffath Mansoor for being the most inspiring women in my life.

My colleague and co-author, Tim, with whom it has been a pleasure working. Thank you for your attention to detail, your patience and your constant 'we can do it attitude' that provided a bedrock of support. A special thanks to Aishwarya (Ash) Narayanan, without whom Tim may not have been in India to write this book with me.

Our colleague Mouli Paul for the lovely illustrations, which helped various elements of the book come alive. Also, a big thank you to our colleague Mehul Manjeshwar who patiently read several drafts, came in early and stayed late at work to help with all the edits before we turned in our manuscript! A shout-out to Michael (Tim's brother) and Tim's mum, Kay, too for their edits across time zones.

Our editor, Tarini Uppal, who pitched this idea to me in 2017, long before I even thought of writing a book! Thank you for being so patient and consistent in following up until we actually started working on it, and once it was on paper, elevating the quality of the guidebook with our star copy editor, Aslesha Kadian.

Finally, I would like to thank my Bare Necessities team for being a constant source of happiness, inspiration and insight throughout this entrepreneurship journey.

—Sahar

NOTES

Chapter 1: Personal Care

1. Vakkalanka, Harshini, 'The Packaging Also Matters', *The Hindu*, 22 May 2018, www.thehindu.com/sci-tech/energyandenvironment/the-packaging-alsomatters/article23958965.ece.

2. Pongrácz, Eva, 'The Environmental Impacts of Packaging', 2007, https://www.researchgate.net/publication/229796182_The_Environmental_Impacts_of_Packaging.

3. 'Skin: Skin Is the Human Body's Largest Organ', National Geographic, 17 January 2017, www.nationalgeographic.com/science/health-and-human-body/human-body/skin/.

4. Ternes, Tracy, 'Your Skin: It Absorbs!', *Down to Earth*, 9 July 2016, www.downtoearth.org/health/general-health/your-skin-it-absorbs.

5. Shata, Doha, 'Miswak: First Toothbrush in History', *Arab News*, 12 August 2013, www.arabnews.com/news/459712.

6. Ibid.

7. Shanbhag, Vagish Kumar L, 'Oil Pulling for Maintaining Oral Hygiene—A Review', *Elsevier Journal of Traditional and Complementary Medicine*, 6 June 2016, www.ncbi.nlm.nih.gov/pmc/articles/PMC5198813/.

8. '7 Reasons to Start Using Coconut Oil Toothpaste Recipes', Carefree Dental, www.carefreedental.com/resources/13-dental-health/130-7-reasons-to-start-using-coconut-oil-toothpaste-recipes.

9. Dry shampoo does not need any water. Simply put a small amount of the mix in your hand, rub it through your hair (or on your scalp) and comb it out. This is really effective if you do not have time for a shower, that is, if you are racing to get to the gym or running late to see a friend for coffee.

10. Plastic comes in various forms. In this book, the types are not elaborated on for the sake of simplicity. However, it is important to understand that there are numerous qualities of plastic. The lower the grade (quality) of plastic the less likelihood there is of the material being recycled. In brief, examples of plastic types and recyclability are: high-density polyethylene (used for plastic crates, lumber, fencing), polyvinyl chloride (used for flooring, mobile home skirting) which is often recyclable but depends on the facilities available. Low-density polyethylene (used for garbage cans and lumber) is even more dependent on the quality and resources of the recycling centres. The unrecyclable plastics include polypropylene (used for ice scrapers, rakes, battery cables), polystyrene or styrofoam (used for insulation, licence plate frames, rulers) and miscellaneous plastics: polycarbonate, polylactide, acrylic, acrylonitrile butadiene, styrene, fiberglass and nylon (used in outdoor decks, moulding and park benches).

    See: Mertes, Alyssa, 'What Are the Different Types of Plastic?', Quality Logo Products, 12 September 2018, www.qualitylogoproducts.com/promo-university/different-types-of-plastic.htm.

11. Not elaborated in the 'Why Is it Harmful?' subsection is the amount of plastic found within common feminine hygiene products. There are multiple layers of plastic in many components of pads, such as the adhesive or wings. Similarly, tampons have plastic throughout, even in the absorbent material itself. Many of these areas were sought after and seen as improvements from the 1960s onwards due to privacy, among other concerns. Key steps to eliminate the amount of plastic (reasons to eliminate plastic are expanded on in the main text) used for these products include menstrual cups and tampons without applicators, which are made from natural materials.

    See: Borunda, Alejandra, 'How Tampons and Pads Became So Unsustainable', National Geographic, 6 September 2019, www.nationalgeographic.com/environment/2019/09/how-tampons-pads-became-unsustainable-story-of-plastic/.

12. There is a risk of Toxic Shock Syndrome (TSS) associated with tampon use. TSS is a type of staph infection that arises when a tampon provides a breeding ground for bacteria. It is scientifically proven that menstrual cups don't change the composition of blood during the time the cup is inside a woman's body. There are no reported cases of TSS in connection with the use of menstrual cups that were invented in the 1920s.

    See: Mitchell, Michael A, et al, 'A Confirmed Case of Toxic Shock Syndrome Associated with the Use of a Menstrual Cup,' *Canadian Journal of Infectious Diseases & Medical Microbiology = Journal Canadien Des Maladies Infectieuses Et De La Microbiologie Medicale*, Pulsus Group Inc., 2015, www.ncbi.nlm.nih.gov/pmc/articles/PMC4556184/ and, Levine, Hallie, 'Menstrual Cup Linked to Toxic Shock Syndrome, New Study Finds', *Consumer Reports*, www.consumerreports.org/women-s-health/menstrual-cups-linked-to-toxic-shock-syndrome/.

13. Lyons, Emily, 'Your Fave Make-up Could Be Harming the Environment', HuffPost Canada, 30 June 2016, www.huffingtonpost.ca/emilylyons/makeuppollution_b_10758282.html

14. The negative impacts of palm oil are not to do with the plant directly. It is a natural product, which if harvested sustainably would not cause the issues that it does. Instead the impacts are felt due to the use of the product. Farmers often burn off crops to stimulate faster regrowth, which is done after large amounts of land clearing (deforestation). These methods of farming lead to air pollution, as witnessed in the 'haze' season in Indonesia, Singapore and Malaysia when the air is filled with hazardous levels of smoke. This can also lead to climate change. Also, the loss of natural forests, replacing it with a monoculture crop in the form of palm oil, has led to habitat loss and the extinction or highly endangered levels of native animals such as the orangutan.
    See: 'Sustainable Palm Oil', *Conservation International*, www.conservation.org/projects/sustainable-palm-oil,
    Maitre, Pascal, et al, 'Palm Oil Is Unavoidable. Can It Be Sustainable?' National Geographic, 7 December 2018, www.nationalgeographic.com/magazine/2018/12/palm-oil-products-borneo-africa-environment-impact/ and,
    Missingham Bruce, et al, 'A Study of Local Livelihoods: Using the Sustainable Livelihoods Approach,' SEACO Asia, 2018, www.seaco.asia/wp-content/uploads/2018/12/2.-Using-The-Sustainable-Livelihoods-Approach.pdf.

15. Anania, Billy, 'Palm Oil Industry Threatens the Future of Borneo's Wildlife', Unreasonable.is, 18 September 2018, https://unreasonablegroup.com/articles/palm-oil-industry-threatens-the-future-of-borneos-wildlife/.

16. 'Origin of Oil Palm' in '2 OIL PALM', Food and Agriculture Organization, www.fao.org/3/y4355e/y4355e03.htm.

17. Anania, Billy, 'Palm Oil Industry Threatens the Future of Borneo's Wildlife', Unreasonable.is, 18 September 2018, https://unreasonablegroup.com/articles/palm-oil-industry-threatens-the-future-of-borneos-wildlife/.

18. Extended Producer Responsibility is a policy approach under which producers are given a significant responsibility (financial and/or physical) for the treatment or disposal of post-consumer products. Assigning such responsibility could, in principle, provide incentives to prevent waste at the source, promote product design for the environment and support the achievement of public recycling and materials management goals. Notably, transparency in the supply chain is a key move forward towards a circular economy.
    See: 'Extended Producer Responsibility', OECD, www.oecd.org/env/tools-evaluation/extendedproducerresponsibility.htm.

19. 'Antibacterial Chemical Disrupts Hormone Activities, Study Finds', ScienceDaily, 8 December 2007, www.sciencedaily.com/releases/2007/12/071207150713.htm.

20.   For more details on chemicals, see Chapter 4.

21.   Ways to 'urge producers to take more responsibility' include (but are not limited to) feedback forms, online reviews, speaking directly to the company, publishing informative posts and blogs and taking part in board meetings (if it is a public company). Some of these areas are discussed in Chapters 6, 7 and 8.

22.   'Pharmaceuticals and Personal Care Products in the Environment: What Are the Big Questions?', National Institute of Environmental Health Sciences, U.S. Department of Health and Human Services, ehp.niehs.nih. gov/doi/full/10.1289/ehp.1104477.

23.   Folk, Emily, 'Common Beauty Industry Environmental Issues', *Conservation Folks*, 1 April 2019, conservationfolks.com/common-beauty-industry-environmental-issues/.

24.   Many products, such as plastics and chemical ingredients, that form the basis of items that are commonly used in the twenty-first century began to be used on an increased scale after innovations were made during wars (not solely in WWII, however, many did occur from the 1950s onwards). This made these products a good business choice. The negative environmental impacts of plastic, for instance, were downplayed by lobbyists, in the USA in particular, and large companies when environmental problems started to be highlighted through data and facts.
See: 'A Brief History of How Plastic Has Changed Our World', YouTube-National Geographic, 2018, www.youtube.com/watch?v=jQdBag_p6kE,
'Why World Has Declared a War against Plastic', *Times of India*, 28 August 2019, timesofindia.indiatimes.com/india/why-world-has-declared-a-war-against-plastic/articleshow/70806066.cms,
Root, Tik, 'Inside the Long War to Protect Plastic', Center for Public Integrity, 16 May 2019, publicintegrity.org/environment/pollution/pushing-plastic/inside-the-long-war-to-protect-plastic/ and,
'History and Future of Plastics', *Science History Institute*, 20 November 2019, www.sciencehistory.org/the-history-and-future-of-plastics.

25.   'The Story of Cosmetics', The Story of Stuff Project, YouTube, 2010, www.youtube.com/watch?v=pfq000AF1i8.

26.   Although policies hold certain business operations to more accountability in the twenty-first century, there are still numerous areas that allow businesses to continue 'business as usual'. This ranges from a lack of laws (specifically to the USA as the largest producer of these products) to companies not needing to list all the ingredients on a product. Additionally, the US Food and Drug Administration does not assess the safety of the ingredients. Instead this area is self-regulated by the businesses producing the products, with compliance to recommendations termed as 'voluntary'. Notably, since 1938 only eight of over 1200 ingredients have been banned.
See: 'The Story of Cosmetics', The Story of Stuff Project, YouTube, 2010, www.youtube.com/watch?v=pfq000AF1i8.

27. More broadly affected is any individual manufacturing the product (often people from lower socioeconomic groups) or people who use the products frequently, such as those who work in beauty salons.
See: 'The Story of Cosmetics', The Story of Stuff Project, YouTube, 2010, www.youtube.com/watch?v=pfq000AF1i8.

28. Systems issues are discussed further in the guidebook while the life cycle of a product will be addressed in Chapter 4.

29. Transparency and responsibility, especially in relation to being an active citizen, are discussed in Chapters 6, 7 and 8.

30. Usmani, Azman, 'There's No Easy Way Out Of the Plastic Mess', Bloomberg Quint, 15 September 2019, www.bloombergquint.com/economy-finance/theres-no-easy-way-out-of-the-plastic-mess.

31. Borunda, Alejandra, 'How Your Toothbrush Became a Part of the Plastic Crisis', 14 June 2019, National Geographic, www.nationalgeographic.com/environment/2019/06/story-of-plastic-toothbrushes/.

32. Toothbrushes have been in use for centuries in varying capacities with many utilizing animal hair, such as hog hair. Similar organic products, such as neem and miswak sticks, have been regularly used in India. Notably though, the design of the modern toothbrush has not changed dramatically from the days when hog hair was attached to a stick in the 1400s by Emperor Hongzhi in China.
See: Borunda, Alejandra, 'How Your Toothbrush Became a Part of the Plastic Crisis', 14 June 2019, National Geographic, www.nationalgeographic.com/environment/2019/06/story-of-plastic-toothbrushes/.

33. A synthetic polymer was first invented in 1869 as a substitute for ivory (caused in part due to the growing popularity of billiards). The creation was seen as a way to relieve people from the social and economic constraints of resource scarcity worldwide. It provided an inexpensive alternative that became more and more available in a modernizing society. By 1907, 'a key breakthrough came . . . the first real synthetic, mass-produced plastic', one that was initially intended for wiring. Throughout the following few decades, new designs and innovation led to new types of plastics being available to the market. The Second World War brought plastics into its own. It became indispensable with observers noting that in 'product after product, market after market, plastics challenged traditional materials and won, taking the place of steel in cars, paper and glass in packaging, and wood in furniture'.
See: 'The History and Future of Plastics,' Science History Institute, 20 December 2016, www.sciencehistory.org/the-history-and-future-of-plastics, 'Plastics: A Story of More Than 100 Years of Innovation,' PlasticsEurope, www.plasticseurope.org/en/about-plastics/what-are-plastics/history, 'Why World Has Declared a War against Plastic', *Times of India*, 28 August 2019, timesofindia.indiatimes.com/india/why-world-has-declared-a-war-against-plastic/articleshow/70806066.cms and,

'The History and Future of Plastics,' Science History Institute, 20 December 2016, www.sciencehistory.org/the-history-and-future-of-plastics.

34. Dyer, Peter, 'Dentistry, the Toothbrush and Sustainability,' British Dental Association, 2019, bda.org/news-centre/blog/Pages/Dentistry-the-toothbrush-and-sustainability.aspx.

35. A plastic, industrially manufactured, disposable sanitary pad requires about 500 to 800 years to decompose. Visually, the impact is that 'thousands of tonnes of disposable sanitary waste is generated every month all over the world and often ends up in the environment due to poor waste management systems.'
    See: Mehrotra, Ayesha, 'Sustainable Menstruation: The Impact of Menstrual Products on the Environment', Medium, 29 April 2018, medium.com/one-future/sustainable-menstruation-theenvironmental-impact-of-menstrual-productseba30e095cda.

36. '7 Ways Your Plastic Toothbrush Is Evil', BamBrush, 7 December 2019, www.bambrushes.com/blogs/news/7-ways-your-plastic-toothbrush-is-evil.

37. Kaur, Banjot, 'No Ban on Single-Use Plastics, to Be Phased out by 2022', *Down To Earth*, 2 October 2019, www.downtoearth.org.in/news/pollution/no-ban-on-single-use-plastics-to-be-phased-out-by-2022-67064.

38. 'Why World Has Declared a War against Plastic', *Times of India*, 28 August 2019, timesofindia.indiatimes.com/india/why-world-has-declared-a-war-against-plastic/articleshow/70806066.cms.

39. Increased focus on these areas can lead to practical, tangible solutions being found, such as replacing one item with another or finding alternate methods. A number of these suggestions are found throughout the chapters in the 'Moving Towards a More Sustainable Lifestyle' subsection.

40. Yashwant, Shailendra, 'Plastic Ban: Single-Use Plastic Has No Place on This Planet', MoneyControl, 11 May 2020, www.moneycontrol.com/news/environment/plastic-ban-single-use-plastic-has-no-place-on-this-planet-4438621.html.

41. Ibid.

42. 'FMCG Industry in India', India Brand Equity Foundation (IBEF), 21 October 2020, www.ibef.org/industry/fmcg.aspx.

43. Ibid.

44. Figures from the Marine Conservation Society reveal that on an average, 4.8 pieces of menstrual waste are found per 100 metres of a beach cleaned. For every 100m of a beach, that amounts to four pads, panty-liners and backing strips, along with at least one tampon and applicator. Astonishingly, on an individual level, 'the average person who menstruates throws away up to 200kg of menstrual products in their lifetime, and this is only one item containing plastic. There are countless others that enter the environment because of ineffective waste system.
    See: 'Plastic Periods: Menstrual Products and Plastic Pollution', Friends of the Earth, 15 October 2018, friendsoftheearth.uk/plastics/plastic-periods-menstrual-products-and-plastic-pollution,

< 298 >

Every day, India produces the equivalent of 9,000 Asian elephants or 86 Boeing 747 planes. To add to this, over half of this comes from the main cities of Delhi, Mumbai, Bengaluru, Chennai and Kolkata and,

Rana, Uday Singh, 'Choking on Our Own Mess: Indians Generate 25,940 Tons of Plastic Waste Every Day,' News18, 19 July 2019, www.news18.com/news/india/choking-on-our-own-mess-indians-generate-25940-tons-of-plastic-waste-every-day-2237567.html.

45. The plastics industry, from the chemical giants forming the building blocks of plastic to companies using the packaging to sell their products, has been waging that war (against banning plastic) for more than thirty years. It has pumped millions of dollars into pro-plastic marketing. All this despite decades of repeated warnings about weak recycling markets and plastic pollution problems.

See: Root, Tik, 'Inside the Long War to Protect Plastic', Center for Public Integrity, 16 May 2019, publicintegrity.org/environment/pollution/pushing-plastic/inside-the-long-war-to-protect-plastic/.

46. The '5 R approach' is a valuable way to describe this concept to the broader public: Refuse what you do not need, Reduce what you do not need, Reuse what you can't reduce, Recycle what you can't reuse, Recover what you can't recycle.

See: 'Waste and Climate Change.' *Waste and Climate Change: Pacific Year of Climate Change 2009*, 2009, www.sprep.org/climate_change/PYCC/documents/ccwaste.pdf.

47. Some information about plastic bans in India in 2019: Adjustments, for a move away from plastic, would need to be made in the context of India, yet by learning from methods that have been effective to date in other locations, a knowledgeable and effective approach can be made to address this systemic issue, which could see a transition away from using materials that are detrimental to the planet, such as single-use plastic.

- France has had a plastic ban since 2016. Within the law there are clauses that specify that 'the replacements of these items will need to be made from biologically sourced materials that can be composted'.
- Rwanda's plastic bag ban is 'not effective just because of strict enforcement but also because of hefty penalties. According to the law, the offenders smuggling plastic bags can face jail time'.
- Sweden has implemented one of the world's best recycling systems that is so effective that 'less than 1 per cent of Sweden's household waste goes into landfill'.
- In Ireland, a dramatic decrease of plastic was seen due to a plastic bag tax in 2002, which saw 'within weeks of its implementation . . . a reduction of 94 per cent in plastic bag use'.
- On a larger scale that parallels India's population, 'China made it illegal for stores (small or big vendors) to give out plastic bags (in 2008) and allowed them to keep any profit they made for themselves. End

result, after two years of the law implementation, usage of plastic bags dropped by a whopping 50 per cent'.

See: Bhatia, Anisha. 'Plastic Ban: What India Can Learn From Other Countries: Plastic Waste.' *NDTV*, 30 June 2017, swachhindia.ndtv.com/plastic-ban-india-can-learn-countries-6161/.

48. 'Why World Has Declared a War against Plastic', *Times of India*, 28 August 2019, timesofindia.indiatimes.com/india/why-world-has-declared-a-war-against-plastic/articleshow/70806066.cms.

49. Items that function within a linear economy: the take, make and dispose process.

See: 'A Circular Economy Differs from a Linear Economy, But How?', Kenniskaarten, kenniskaarten.hetgroenebrein.nl/en/knowledge-map-circular-economy/how-is-a-circular-economy-different-from-a-linear-economy/.

50. US Department of Commerce, and National Oceanic and Atmospheric Administration (NOAA), 'What Are Microplastics?', NOAA's National Ocean Service, 13 April 2016, oceanservice.noaa.gov/facts/microplastics.html.

51. Drahl, Carmen, 'What You Need To Know About Microbeads, The Banned Bath Product Ingredients', *Forbes*, 9 January 2016, www.forbes.com/sites/carmendrahl/2016/01/09/what-you-need-to-know-about-microbeads-the-banned-bath-product-ingredients/#291e13d47a33.

52. A prime example of addressing these issues is found in the actions of Slovakia's President Zuzana Caputova, before she took over as President. For more than a decade, she addressed major pollution concerns in the town that she lived in, Pezinok, which were causing harm to people and the environment. For more than fourteen years she waged a legal and administrative battle to shut down the town's landfill, which contained the municipal waste of many towns, including the capital, Bratislava. Using scientific data, she was able to prove a link between the presence of waste and high levels of illness among the people of Pezinok. In protesting the landfill, she led the biggest grassroots movement in the country since the Velvet Revolution (1989). Success came in 2013 when the Slovakian Supreme Court cancelled the landfill's license.

See: 'Meet Slovakia's New President, the Erin Brockovich of Central Europe', *Leaders League*, 20 May 2019, www.leadersleague.com/en/news/meet-slovakia-s-new-president-the-erin-brockovich-of-central-europe.

53. S., Lekshmi Priya, 'A Step-By-Step Guide to Making Your Own Eco-Friendly Sanitary Pads. It's Really Simple!', The Better India, 24 August 2017, www.thebetterindia.com/112835/diy-cloth-based-sanitary-pads/.

Chapter 2: Closet

1. Gandhi, Shweta, et al, 'Ever Wondered What Happens to Your Clothes after You Discard Them?' *Vogue India*, 1 March 2019, www.vogue.in/content/ethical-fashion-what-happens-clothes-discard.

2. Gandhi, Shweta, 'Ever Wondered What Happens to Your Clothes after You Discard Them?', *Vogue India*, 2 March 2019, www.vogue.in/content/ethical-fashion-what-happens-clothes-discard.

3. Thomas, Maria, 'There Are 100 Ways to Wear a Sari-How Many Do You Know?', Quartz India, 21 October 2016, qz.com/india/812115/there-are-100-ways-to-wear-a-sari-how-many-do-you-know/.

4. Maheshwari, Sapna, 'Forever 21 Bankruptcy Signals a Shift in Consumer Tastes,' *New York Times*, 29 September 2019, www.nytimes.com/2019/09/29/business/forever-21-bankruptcy.html.

5. 'Forget Shopping! You've Got To Swap Clothes At This Coolest Clothing Exchange: LBB', *LBB, Hyderabad*, lbb.in/hyderabad/switcheroo-clothing-exchange/.

6. For more, visit www.globalshapers.org/.

7. Majumdar, Meghna and Narrain, Aparna, 'Clothes Swaps and Second-Hand Sales Become Trendy in Indian Cities,' *The Hindu*, 28 October 2019, www.thehindu.com/life-and-style/fashion/clothes-swaps-and-second-hand-sales-become-trendy-in-indian-cities/article29786995.ece.

8. Ibid.

9. Ibid.

10. For more, visit www.apparentclub.com/?fbclid=IwAR232U_8Td7mmHfA3 6PdiAZrkAhWbQA-zKDyf4Hp3vLHtIXbZoUrpDIzvAQ.

11. For their Facebook page, visit www.facebook.com/apparentclub/?ref=page_internal.

12. Majumdar, Meghna, and Narrain, Aparna, 'Clothes Swaps and Second-Hand Sales Become Trendy in Indian Cities,' *The Hindu*, 28 October 2019, www.thehindu.com/life-and-style/fashion/clothes-swaps-and-second-hand-sales-become-trendy-in-indian-cities/article29786995.ece.

13. Ibid.

14. 'Fashion Industry Waste Statistics,' E D G E, 8 January 2020, edgexpo.com/fashion-industry-waste-statistics/.

15. Joy, Annamma; Sherry, John F. Jr.; Venkatesh, Alladi; Wang, Jeff; Chan, Ricky, 'Fast Fashion, Sustainability, and the Ethical Appeal of Luxury Brands, *Fashion Theory*, 2012, www3.nd.edu/~jsherry/pdf/2012/FastFashionSustainability.pdf.

16. 'Rana Plaza', Clean Clothes Campaign, 1 September 2019, cleanclothes.org/campaigns/past/rana-plaza.

17. A case in point of this scenario is that, in late 2014, a report highlighted that only a little over a third of the compensation had been paid to the victims despite the horrific nature of the incident and the global recognition that this was a tragedy that should not have occurred and should not happen ever again.
See: O'Connor, Clare, 'These Retailers Involved in Bangladesh Factory Disaster Have Yet to Compensate Victims,' *Forbes*, 26 April 2014, www.forbes.com/sites/clareoconnor/2014/04/26/these-retailers-

involved-in-bangladesh-factory-disaster-have-yet-to-compensate-victims/#28285a3c211b.

18. Bangladesh has a low minimum wage requirement, cheap land and low operation costs. As a result, these conditions have allowed corporations to take advantage of the situation and employees in unethical ways.

19. Ecosystems and lifestyle analysis will be discussed in detail in Chapter 4.

20. Cline, Elizabeth L., *Overdressed: The Shockingly High Cost of Cheap Fashion*, Portfolio (Penguin), 2013.

21. 'The Impact of COVID-19 on the People Who Make Our Clothes', Fashion Revolution, 11 May 2020, www.fashionrevolution.org/the-impact-of-covid-19-on-the-people-who-make-our-clothes/.

22. 'Brands Must Urgently Take Steps to Minimise Impact of the Coronavirus on Garment Workers' Health and Livelihoods',Clean Clothes Campaign, 18 March 2020, cleanclothes.org/news/2020/brands-must-urgently-take-steps-to-minimise-impact-of-the-coronavirus-on-garment-workers-health-and-livelihoods.

23. Coronavirus disease (COVID-19) is an infectious disease caused by a newly discovered coronavirus.
See: 'Coronavirus', World Health Organization, www.who.int/health-topics/coronavirus#tab=tab_1.

24. 'Timeline – How Coronavirus Is Impacting the Global Apparel Industry – FREE TO READ', Just-Style, 6 November 2020, www.just-style.com/news/timeline-how-coronavirus-is-impacting-the-global-apparel-industry-free-to-read_id138313.aspx.

25. 'The Impact of COVID-19 on the People Who Make Our Clothes', Fashion Revolution, 11 May 2020, www.fashionrevolution.org/the-impact-of-covid-19-on-the-people-who-make-our-clothes/.

26. Estimate from March 2020.

27. 'The Impact of COVID-19 on the People Who Make Our Clothes', Fashion Revolution, 11 May 2020, www.fashionrevolution.org/the-impact-of-covid-19-on-the-people-who-make-our-clothes/.

28. Lin, Jesse; Lanng, Christian, 'Here's How Global Supply Chains Will Change after COVID-19', World Economic Forum, 6 May 2020, www.weforum.org/agenda/2020/05/this-is-what-global-supply-chains-will-look-like-after-covid-19?fbclid=IwAR19pp0r-ZwtiG2LRJDdY-H2tUQ1i-1lnaG780i4LaQsuIOqgic8i4iSYps.

29. 'Covid-19: Impact on Brands and Workers in Garment Supply Chains', Fair Wear, www.fairwear.org/covid-19-dossier/covid-19-guidance/.

30. Jones, Katie, 'How COVID-19 Consumer Spending Is Impacting Industries', Visual Capitalist, 22 April 2020, www.visualcapitalist.com/consumer-spending-impacting-industries/.

31. Kumar, Neeraj; Haydon, Danny, 'Industries Most and Least Impacted by COVID-19 from a Probability of Default Perspective – March 2020 Update', S&P Global Market Intelligence, 7 April 2020,

www.spglobal.com/marketintelligence/en/news-insights/blog/industries-most-and-least-impacted-by-covid-19-from-a-probability-of-default-perspective-march-2020-update.

32. In India, the textile industry employs the second-highest number of people after agriculture. Impacting the lives of these employees creates has a significant effect on an enormous number of people.
     See: Sharma, Seema, 'New Textile Policy to Boost Employment', Employment News, employmentnews.gov.in/NewEmp/MoreContentNew.aspx?n=Editorial&k=70.

33. 'The "Reset" It Needed: How Coronavirus Is Changing the Fashion Industry Forever', ABC News, 29 April 2020, www.abc.net.au/news/2020-04-28/coronavirus-covid19-australia-fashion-industry-adapts/12189924.

34. McFall-Johnsen, Morgan, 'The Fashion Industry Emits More Carbon than International Flights and Maritime Shipping Combined. Here Are the Biggest Ways It Impacts the Planet,' *Business Insider Australia*, 17 October 2019, www.businessinsider.com.au/fast-fashion-environmental-impact-pollution-emissions-waste-water-2019-10?r=US&IR=T.

35. 'Elizabeth Cline: Author of Overdressed: The Shockingly High Cost of Fast Fashion', Make.Good, www.makegood.world/interview-elizabeth-cline.

36. Global consumerism has created fifty-two micro-seasons that have moved the fashion industry into one of the most polluting industries globally. The impact itself is felt all the way through the supply chain: initially crops need to be grown (adding pesticides to the environment, particularly for cotton, which harms the planet and the farmer), material needs to be bound together by workers who are given high targets at minimal personal reward, consumers are asked to buy for each new party or event because advertising suggests that they should do so. Finally, the landfills, rivers and oceans are polluted by the excess material that was disregarded by a consumer attempting to fit into a micro-season trend.
     See: Azevedo, Andrea, 'The Impact of the 52 Micro-Seasons on the Environment', Medium, 2 April 2018, medium.com/@andreaazevedo_32670/the-effects-of-the-52-micro-seasons-on-the-environment-edd87951b74f.

37. Souchet, Francois and Ellen MacArthur Foundation, 'Fashion Has a Huge Waste Problem. Here's How It Can Change', World Economic Forum, www.weforum.org/agenda/2019/02/how-the-circular-economy-is-redesigning-fashions-future/

38. Ibid.

39. 'Worn Wear: Better Than New', Patagonia Outdoor Clothing & Gear, www.patagonia.com/worn-wear.html.

40. 'Did You Know Less Than 1% of Old Clothing Becomes New Clothes? Make a Circular Economy for Fashion', Ellen MacArthur Foundation, YouTube, 21 November 2018, www.youtube.com/watch?v=3iKHr-JnWYA&feature=youtu.be.

41. 'A New Textiles Economy', Ellen MacArthur Foundation, www. ellenmacarthurfoundation.org/assets/downloads/publications/A-New-Textiles-Economy_Full-Report_Updated_1-12-17.pdf.

42. Fashion Revolution is a global movement that runs all year long with a specific 'Fashion Revolution Week' organized annually in the third week of April, to commemorate the Rana Plaza tragedy. They celebrate fashion as a positive influence while scrutinizing industry practices and raising awareness of the fashion industry's most pressing issues. They aim to show that change is possible and encourage those who are on a journey to create a more ethical and sustainable future for fashion. They look at the issues as part of a system where areas need to be addressed to create a circular economy in all areas of fashion. For more, visit www.fashionrevolution.org/about/.

43. Drew, Deborah; Reichart, Elizabeth, 'These Are the Economic, Social and Environmental Impacts of Fast Fashion', World Economic Forum, 11 January 2019, www.weforum.org/agenda/2019/01/by-the-numbers-the-economic-social-and-environmental-impacts-of-fast-fashion.

44. Nabity, Steve, 'How to Make a T-Shirt Quilt in 4 Easy Steps', MemoryStitch, 31 May 2019, memorystitch.com/blogs/news/t-shirt-quilts-in-5-easy-steps.

45. Giordano, Medea, 'How to Make a CDC-Approved Cloth Face Mask', Wired, 29 July 2020, www.wired.com/story/how-to-make-a-cloth-face-mask/.

46. Ibid.

47. Ibid.

48. Wills, Melissa, 'DIY Recycled Towel Bathmat', BuzzFeed, www.buzzfeed. com/bfmp/videos/31089.

49. 'No Mess, No Stink, No Leaky Diapers! White Cloth Nappies for Toddlers', Bumpadum, www.bumpadum.com/pages/about-us.

50. 'Cloth Diapers India: Reusable Cloth Diapers India', Superbottoms, www. superbottoms.com/.

51. 'The Things We Do Make a Difference Every Day', Just Little Changes, 19 December 2019, justlittlechanges.com/.

Chapter 3: Kitchen

1. Felder, Steve, 'How to Turn India's Food Waste Problem into Opportunity', *Forbes* (India), 28 January 2019, www.forbesindia.com/blog/technology/how-to-turn-indias-food-waste-problem-into-opportunity/.

2. Ibid.

3. 'Urban Development Series-Chapter 5', Urban Development Series-Knowledge Papers Chapter 5 Waste Composition, World Bank, siteresources.worldbank.org/INTURBANDEVELOPMENT/Resources/336387-1334852610766/Chap5.pdf.

4. Ammal, S. Meenakshi and Chawla, Ashish, *The Best of Samaithu Paar: The Classic Guide to Tamil Cuisine*, Viking, 2001.

5. 'What Is the Problem with Cling Film?', BeeBee Wraps, beebeewraps.com/blogs/news/what-isthe-problem-with-cling-film

6. Gibbens, Sarah, 'The Sticky Problem of Plastic Wrap', National Geographic, 12 July 2019, www.nationalgeographic.com/environment/2019/07/story-of-plastic-sticky-problem-of-plastic-wrap/

7. Ibid.

8. Ibid.

9. It should be noted that this cost does not take into account costs outside of running that specific business, i.e. it does not take into account the environmental costs due to the linear system of production and consumption.

10. Lerner, Sharon, 'Coca-Cola Named Most Polluting Brand in Global Plastic Waste Audit', Intercept, 23 October 2019, theintercept.com/2019/10/23/coca-cola-plastic-waste-pollution/.

11. '10 Facts About Single-Use Plastic Bags', Centre for Biological Diversity, www.biologicaldiversity.org/programs/population_and_sustainability/sustainability/plastic_bag_facts.html.

12. Excell, Carole, '127 Countries Now Regulate Plastic Bags. Why Aren't We Seeing Less Pollution?', *World Resources Institute*, 11 March 2019, www.wri.org/blog/2019/03/127-countries-now-regulate-plastic-bags-why-arent-we-seeing-less-pollution.

13. Khadka, Shyam, 'Reducing Food Waste Vital for India's Food Security-UN India', United Nations, 14 March 2017, in.one.un.org/reducing-food-waste-vital-indias-food-security/.

14. Food losses and waste amounts to roughly US$ 680 billion in industrialized countries and US$ 310 billion in developing countries. Industrialized and developing countries dissipate roughly the same quantities of food—respectively 670 and 630 million tonnes. Fruits and vegetables, plus roots and tubers, have the highest wastage rates of any food. Global quantitative food losses and waste per year are roughly 30 per cent for cereals, 40–50 per cent for root crops, fruits and vegetables, 20 per cent for oil seeds, meat and dairy, plus 35 per cent for fish.

15. The Sustainable Development Goals (SDGs), also known as the Global Goals, were adopted by all United Nations member states in 2015 as a universal call to action to end poverty, protect the planet and ensure that all people enjoy peace and prosperity by 2030. The seventeen SDGs are integrated—that is, they recognize that action in one area will affect outcomes in others, and that development must balance social, economic and environmental sustainability.
See: 'Sustainable Development Goals', UNDP, www.undp.org/content/undp/en/home/sustainable-development-goals.html.

16. 'In February 2016, France became the first country in the world to prohibit supermarkets from throwing away unused food through a unanimously passed legislation. Now, supermarkets of a certain size must donate unused food or face a fine. Other policies require schools to teach students about

food sustainability, companies to report food waste statistics in environmental reports and restaurants to make take-out bags available.'

See: Hinckley, Story, 'France Was the First Country to Ban Supermarkets from Throwing Away Unused Food-and the World Is Taking Notice', *Business Insider*, 7 January 2018, www.businessinsider.com/how-france-became-a-global-leader-in-curbing-food-waste-2018-1?IR=T.

17. For more about Food for Soul, visit www.foodforsoul.it/.

18. 'The Food Chain, The Real Junk Food', BBC World Service, 24 June 2018, www.bbc.co.uk/programmes/w3cswpmm.

19. 'Fight Climate Change by Preventing Food Waste', World Wildlife Fund, www.worldwildlife.org/stories/fight-climate-change-by-preventing-food-waste.

20. 'Food Wastage in India, And What You Can Do About It', *CSR Journal*, 31 October 2018, thecsrjournal.in/food-wastage-in-india-a-serious-concern/.

21. 'How to Turn India's Food Waste Problem into Opportunity', *Forbes India*, 28 January 2019, www.forbesindia.com/blog/technology/how-to-turn-indias-food-waste-problem-into-opportunity/.

22. Morrison, Leslie, 'The Benefits of Eating Seasonal Produce and Our Favorite Spring Fruits and Vegetables', Deliciously Plated, 27 February 2019, deliciouslyplated.com/food-articles/the-benefits-of-eating-seasonal-produce/.

23. Climate change is discussed at more length in Chapter 8.

24. Flanagan, Katie, 'Is the World on Track to Cut Food Loss and Waste in Half by 2030?', World Resources Institute, 23 September 2019, www.wri.org/blog/2019/09/world-track-cut-food-loss-and-waste-half-2030.

25. 'Eating Seasonally-Why It's Good for the Planet, Your Wallet and Your Waistline', Sanitarium Health Food Company, www.sanitarium.com.au/health-nutrition/nutrition/eating-seasonally.

26. 'Food Wastage in India, And What You Can Do About It', *CSR Journal*, 31 October 2018, thecsrjournal.in/food-wastage-in-india-a-serious-concern/.

27. Ibid.

28. 'Solar Dryers Reduce Farmers' Food Loss in India', UN Environment, www.unenvironment.org/news-and-stories/story/solar-dryers-reduce-farmers-food-loss-india.

29. 'Porpeanglife.com : Integrated Agriculture', AdAge, 25 March 2008, adage.com/creativity/work/integrated-agriculture/1868

30. Watts, Ben, 'The Dangers of Monoculture Farming', Challenge Advisory, 8 October 2018, www.challenge.org/knowledgeitems/the-dangers-of-monoculture-farming/.

31. Ehrmann, Jurgen and Ritz, Karl, 'Plant: Soil Interactions in Temperate Multi-Cropping Production Systems.' *Plant and Soil*, Springer International Publishing, 1 January 1970, link.springer.com/article/10.1007/s11104-013-1921-8.

32. There are around 500 million small-holder farmers in the world and they produce up to 80 per cent of the food consumed in Africa and Asia. They are net buyers of food and very vulnerable to increases in food prices. As a group, they are among the poorest and most marginalized in the world. They are also stewards of increasingly scarce natural resources and are on the frontline of dealing with the impacts of climate change. Smallholders therefore play a critical role in addressing the challenges of food security, poverty and climate change.

33. An example of small-holder vulnerability in India: 'India's agriculture needs to become climate resilient, as the risks for small-scale farmers become more dangerous with each passing year. As both a contributor to climate change and a victim of its impacts, agriculture needs to become climate resilient. This direct connection between climate change and agriculture is perhaps nowhere more apparent than in India, where recent research has shown climate change as the key contributing factor to the suicides of more than 60,000 farmers. This shocking number reveals the deep social and psychological impacts of climate on small-holder farmers and agricultural workers, who form the majority of the poor and hungry, and calls for better tools, policies and programmes to address the growing threat.'
See: 'Farmer Suicides: A Call to Climate Action for India', ReliefWeb, reliefweb.int/report/india/farmer-suicides-call-climate-action-india.

34. Jim Robbins, et al. 'With New Perennial Grain, a Step Forward for Eco-Friendly Agriculture.' *Yale E360*, e360.yale.edu/features/with-new-perennial-grain-a-step-forward-for-eco-friendly-agriculture

35. Manhas, Karn, 'Why Is No One Talking about Agriculture As a Solution to Climate Change?', AgFunderNews, 28 May 2020, agfundernews.com/why-is-no-one-talking-about-agriculture-as-a-solution-to-climate-change.html.

36. This statistic focuses on reports from Ghana and Côte d'Ivoire, which are two of the world's largest cocoa-producing countries. The countries grew and exported 2.3 million metric tonnes (52 per cent) of the world's cocoa in 2016.

37. Stone, Emily, 'Removing Child Labor, Deforestation, and Poverty from the Cost of Chocolate', Unreasonable, 5 September 2018, unreasonable.is/removing-child-labor-deforestation-poverty-from-cost-of-chocolate/.

38. 'In the country's capital, Seoul, 6000 automated bins equipped with scales and Radio Frequency Identification (RFID) weigh food waste as it is deposited and charge residents using an ID card. The pay-as-you-recycle machines have reduced food waste in the city by 47,000 tonnes in six years, according to city officials. Residents are urged to reduce the weight of the waste they deposit by removing moisture first. Not only does this cut the charges they pay—food waste is around 80 per cent moisture—but it also saved the city US $8.4 million in collection charges over the same period. Waste collected using the biodegradable bag scheme is squeezed at the processing plant to remove moisture, which is used to create biogas and bio

oil. Dry waste is turned into fertilizer that is, in turn, helping to drive the country's burgeoning urban farm movement.'
See: Broom, Douglas, 'South Korea Once Recycled 2 per cent of Its Food Waste. Now It Recycles 95 per cent', World Economic Forum, 12 April 2019, www.weforum.org/agenda/2019/04/south-korea-recycling-food-waste/.

39. Douglas, 'South Korea Once Recycled 2 per cent of Its Food Waste. Now It Recycles 95 per cent', World Economic Forum, www.weforum.org/agenda/2019/04/south-korea-recycling-food-waste/.

40. Woo, Ryan and Seun, Thomas, 'Billions of Cockroaches Are Being Farmed in China to Tackle Food Waste', World Economic Forum, 11 December 2018, www.weforum.org/agenda/2018/12/bug-business-cockroaches-corralled-by-the-millions-in-china-to-crunch-waste.

41. Funnell, Antony, 'The People Turning Farms into Bio-Reactors in a Bid to Save the Planet', ABC News, 9 November 2019, www.abc.net.au/news/2019-11-09/controlled-environmental-agriculture-urban-farming/11672818.

42. Although reducing, recycling and reusing are better options for long-term environmental sustainability, this solution addresses the current issue that globally landfill sites produce 10 per cent of the world's methane.

43. Gray, Alex, 'This African City Is Turning a Mountain of Trash into Energy', World Economic Forum, 9 May 2018, www.weforum.org/agenda/2018/05/addis-ababa-reppie-trash-into-energy/?utm_source=Facebook%2BVideos&utm_medium=Facebook%2BVideos&utm_campaign=Facebook%2BVideo%2BBlogs.

44. 'How to Turn India's Food Waste Problem into Opportunity', Forbes India, 28 January 2019, www.forbesindia.com/blog/technology/how-to-turn-indias-food-waste-problem-into-opportunity/.

45. Plan ahead and buy only what you need, use your freezer to store products, be creative with leftovers, blend, bake or boil fruits and vegetables when the quality lessens and speak to others about the need to reduce food waste.

46. 'Fight Climate Change by Preventing Food Waste', World Wildlife Fund, www.worldwildlife.org/stories/fight-climate-change-by-preventing-food-waste.

47. 'Food Waste', World Wildlife Fund, www.worldwildlife.org/initiatives/food-waste.

48. Flanagan, Katie, 'Is the World on Track to Cut Food Loss and Waste in Half by 2030?', World Resources Institute, 23 September 2019, www.wri.org/blog/2019/09/world-track-cut-food-loss-and-waste-half-2030

49. Dixon, John and Gulliver, Aidan with Gibbon, David, Introduction to 'Farming Systems and Poverty', FAO and World Bank, 2001, www.fao.org/3/ac349e/ac349e03.htm.

50. Systems thinking will be discussed further in Chapter 4. However, briefly, while discussing food waste and the interconnected nature of all parts of the process to highlight waste throughout the entire system: if you were to waste

a kilogram of wheat, you would waste 1500 litres of water. Had it been rice, it would be 3500 litres of this precious liquid.

51. 'Fight Climate Change by Preventing Food Waste', World Wildlife Fund, www.worldwildlife.org/stories/fight-climate-change-by-preventing-food-waste.

52. 'Solutions for Reducing Food Loss and Ensuring Sustainable Fishing Livelihoods', FAO, www.fao.org/in-action/bycatch-solutions-latin-america-caribbean/en/.

53. Ouya, Daisy, 'Why Food Waste Is a Concern for Tropical Forest Conservation', Agroforestry World, 16 October 2015, blog.worldagroforestry.org/index. php/2015/10/16/why-food-waste-is-a-concern-for-tropical-forests-conservation/.

54. 'Disasters Causing Billions in Agricultural Losses, with Drought Leading the Way', FAO, 15 March 2018, www.fao.org/geneva/news/detail/en/c/1109577/.

55. Reefs around Australia had been dredged to the brink of extinction due to overfishing of shellfish, including clams, oysters and fish. An initiative between the government and a global NGO, The Nature Conservancy, has seen the regrowth of these reefs through careful planning and sustainable use of the area. This has notable wide-reaching effects with an oyster able to filter a bath water amount of water in a day, thus improving the environment for all living creatures in that location.

See: Knight, Ben, 'After Being Fished to near-Extinction Last Century, Australia's Lost Shellfish Reefs Are Roaring Back to Life', ABC News, 9 November 2019, www.abc.net.au/news/2019-11-09/project-brings-near-dead-reefs-back-to-life-across-australia/11681904.

56. Food systems and farming are exceptionally diverse, ranging from large-scale production of products to small, independent farms. Worldwide, agriculture is part of the solution to a sustainable food system that can sustainably feed a growing population, but traditional systems need to be assessed with objective and accurate data and facts about how to minimize environmental impact while providing sustenance for people. A case of where objectivity in evaluating food systems is needed is found in the focus on reducing types of farming, especially meat, to reduce environmental impact. Poignantly, research suggests that impact is made to the environment by farming both plant and animal products, and that the cessation of one type of farming will not solve contemporary issues (animal products provide sustenance to people who may not have plant-based resources at hand in certain locations in the world). Instead, the key move forward is managing land and resources in an environmentally friendly fashion that minimizes impact. Note that in the two largest meat-eating countries, the USA and Australia, a more balanced diet, i.e. reducing the amount of animal products, has been recommended. However, due to the outlying percentage consumption that these two countries represent compared to global averages, an objective

global perspective in developing a new global food system (that suits certain locations while understanding broader effects) holds the best potential of finding a sustainable system.

57. Systems and life cycles of products are discussed in more depth in Chapter 4.

58. For more, visit www.copperandcloves.com/.

59. 'Feeding India: Our Work', Zomato Feeding India, www.feedingindia.org/ourwork.

60. 'The Robin Hood Army', 17 May 2020, robinhoodarmy.com/.

61. Ibid.

62. 'Start a Green Spot', *SwachaGraha*, www.swachagraha.in/.

Chapter 4: Home Care

1. Bagai, Eric, 'Chemical Water Pollution Caused by Every Day Detergents', Sciencing, 2 March 2019, sciencing.com/chemical-pollution-caused-day-detergents-6664097.html.

2. Ibid.

3. 'Soap lather, which suspends dirt by creating greater surface tension in water, traps dirt for easy removal through rinsing. But we don't require a ton of lather to get the job done.'
See: 'The Real Dirt on Soap Lather', Environmental Enthusiast, 2 August 2011, www.environmentalenthusiast.com/2011/08/the-real-dirt-on-soap-lather/.

4. 'Vani Murthy', YouTube, www.youtube.com/channel/UCzdSzEfw_kGOQyCe19MGgFw.

5. Roux, Erene, 'Watch: Here's Why Kitchen Sponges Are Bad for the Environment [Video]', South African, 19 November 2018, www.thesouthafrican.com/videos/watch-heres-why-kitchen-sponges-are-bad-for-the-environment-video/.

6. Jacobson, Julie, '15 Tips to Make Your Home More Green', Redfin Blog, 15 July 2019, www.redfin.com/blog/15-tips-to-make-your-home-more-green/.

7. '10 Ways to Make Your Home Eco-Friendly', Blue and Green Tomorrow, 4 February 2020, blueandgreentomorrow.com/features/10-ways-make-home-eco-friendly/.

8. Found in fragranced products such as air fresheners, dish soap and toilet paper.

9. Found in dry-cleaning solutions, spot removers, and carpet and upholstery cleaners.

10. Found in most liquid detergents and hand soaps that are labelled 'antibacterial'.

11. Found in fabric softener liquids and sheets, and most household cleaners labelled 'antibacterial'.

12. Found in window, kitchen and multipurpose cleaners.

13. Found in polishing agents for bathroom fixtures, sinks and jewellery. Also found in glass cleaners.

14. Found in scouring powders, toilet bowl cleaners, mildew removers, laundry whiteners.

15. Found in oven cleaners and drain openers.

16. Sholl, Jessie, '8 Hidden Toxins: What's Lurking in Your Cleaning Products?', Experience Life, October 2011, experiencelife.com/article/8-hidden-toxins-whats-lurking-in-your-cleaning-products/.

17. Ibid.

18. Ibid.

19. This has been discussed in detail in Chapter 1.

20. 'The Story of Cosmetics', The Story of Stuff Project, YouTube, 20 July 2010, www.youtube.com/watch?v=pfq000AF1i8.

21. Acaroglu, Leyla, 'A Guide to Life Cycle Thinking', Disruptive Design, 6 March 2018, medium.com/disruptive-design/a-guide-to-life-cycle-thinking-b762ab49bce3

22. The scientific process of understanding what impacts occur as a result of the materials that move through the economy is called a life cycle assessment/analysis.

23. Roux, Erene, 'Watch: Here's Why Kitchen Sponges Are Bad for the Environment [Video]', South African, 19 November 2018, www.thesouthafrican.com/videos/watch-heres-why-kitchen-sponges-are-bad-for-the-environment-video/.

24. Ibid.

25. The World Bank has highlighted the need to manage all waste to a greater level, and that low-income countries mismanage 90 per cent of the waste generated. Further, the amount of waste is expected to increase by 70 per cent globally to 3.4 billion tonnes annually if current practices continue. Increasing recycling and limiting plastic are important steps forward in order to avoid these scenarios and those that can be practiced at home. Notably, this is only one step throughout a life cycle, which is why multiple stakeholders at all stages of a product's life need to be involved. Transparency in design, production, distribution, use and post-consumer life is helpful in establishing better practices.
See: 'What a Waste 2.0', World Bank, 2018, www.worldbank.org/content/dam/infographics/780xany/2018/sep/What-A-Waste-Infographic-780p.png and,
Hares, Sophie, 'Global Waste Could Increase by 70 per cent by 2050, According to the World Bank', World Economic Forum, www.weforum.org/agenda/2018/09/world-waste-could-grow-70-percent-as-cities-boom-warns-world-bank/.

26. To limit household waste, new innovations need to be found such as that led by The Loop Alliance, which is a collaboration between a global recycling and waste management company, TerraCycle, and logistics company UPS.

The aim is to promote responsible consumption and eliminate waste by introducing a new way for consumers to purchase, enjoy and recycle their favourite products. Instead of relying on single-use packaging, it delivers products to the consumer's doorsteps in durable packaging that is collected, cleaned, refilled and reused. Forward thinking in this fashion can have a profound impact on how you live at home and interact with the world around you. It also promotes knowledge of the life cycle of the product, which is a vital step in a move towards a circular economy.

See: Rooney, Katharine Rooney, 'Plastic Packaging Problem: Five Innovative Ideas', World Economic Forum, 17 October 2019, www.weforum.org/agenda/2019/10/plastic-packaging-environment-design-loop/.

'The Loop Alliance Plans to Eliminate Plastic Waste and Save the Planet. You Can Too', World Economic Forum, 26 March 2019, www.weforum.org/our-impact/the-loop-alliance-plans-to-eliminate-plastic-waste-and-save-the-planet-you-can-too.

27. Banwell, Eleanor, 'Repel Microbes from Surfaces', AskNature, 24 March 2020, asknature.org/collections/repel-microbes-from-surfaces/

28. 'StoColor Lotusan Paint', AskNature, 5 November 2020, asknature.org/idea/stocolor-lotusan-paint/.

29. Approximately '60 per cent of material made into clothing is made of plastic. And as many as 729,000 fibres could be released from a single 6 kg laundry load of synthetic materials.' To counter this, Japanese scientists are trialling a 'bulk acoustic wave system' to trap the micro-plastic in the machine so that it cannot escape into the oceans.

See: Fleming, Sean, 'These Scientists Are Using Sound Waves to Filter Plastic Fibres from Laundry Wastewater', World Economic Forum, 2 January 2020, www.weforum.org/agenda/2020/01/acoustic-waves-laundry-plastic-pollution.

30. 'Laundry washed with detergent produces, on average, 86 per cent more microfibers than laundry washed with pure water. This is most likely due to the general mechanics of how detergent works—by loosening fabric's fibres for ease of cleaning. During the cleaning process, many small bits of fibre come off of the fabric and drain from the washing machine with the wastewater. So much so that synthetic microfibers make up 35 per cent of plastic waste.'

See: Deroberts, Nicole, 'Washing Laundry May Be an Underappreciated Source of Microplastic Pollution', Phys.org, 22 August 2019, phys.org/news/2019-08-laundry-underappreciated-source-microplastic-pollution.html.

31. 'Scientists Discovered a Simple Laundry Tweak That Can Cut Down on Ocean Pollution', Curiosity, curiosity.com/topics/scientists-discovered-a-simple-laundry-tweak-that-can-cut-down-on-ocean-pollution-curiosity?utm_source=androidapp.

32. 'The Microbead-Free Waters Act of 2015 aimed to reduce micro-plastic pollution by prohibiting the addition of plastic microbeads in personal-care

products such as face wash, shampoo and toothpaste. However, this law did not regulate microbeads for industrial purposes. A common item that falls under the industrial category is laundry detergent.'
See: Deroberts, Nicole, 'Washing Laundry May Be an Underappreciated Source of Microplastic Pollution', Phys.org, 22 August 2019, phys.org/news/2019-08-laundry-underappreciated-source-microplastic-pollution.html.

33. Deroberts, 'Washing Laundry May Be an Underappreciated Source of Microplastic Pollution', phys.org/news/2019-08-laundry-underappreciated-source-microplastic-pollution.html.Ibid.

34. Pandey, Arundhati, 'What India Can Teach the World about Sustainability', World Economic Forum, 2 October 2017, www.weforum.org/agenda/2017/10/what-india-can-teach-the-world-about-sustainability/.

35. 'How to Turn Your Apartment into a Sustainable Home', Sustainability for All, www.activesustainability.com/sustainable-life/how-to-turn-your-apartment-into-a-sustainable-home/.

36. A green roof or green building is a modern design technique of utilizing rooftops, or the sides of walls, to add vegetation. This has the following benefits: provides a rainwater buffer, purifies the air, reduces ambient temperature, increases solar panel efficiency, reduces ambient noise outside and extends the lifespan of the roof, adds value to the building, increases biodiversity, creates a fire-resistant layer, increases the feeling of well-being, offers healing environment when it is located at a hospital, adds to social interaction, reduces vandalism is low-maintenance. This is discussed further in Chapter 8.
See: 'Benefits of a Green Roof', Sempergreen, www.sempergreen.com/en/solutions/green-roofs/green-roof-benefits.

37. Reduction in costs of renewable energy products, solar power for instance, is starting to change practices in this regard. See: Sun, Xiaojing, 'Solar Technology Got Cheaper and Better in the 2010s. Now What?', Greentech Media, 17 December 2019, www.greentechmedia.com/articles/read/solar-pv-has-become-cheaper-and-better-in-the-2010s-now-what.

38. 'How to Turn Your Apartment into a Sustainable Home', Sustainability for All, www.activesustainability.com/sustainable-life/how-to-turn-your-apartment-into-a-sustainable-home/.

39. Vellapally, Sunita, '7 Sustainable Home Design Ideas for Indian Homes', Homify, 24 October 2019, www.homify.in/ideabooks/6326971/7-sustainable-home-design-ideas-for-indian-homes.

40. There are many things that you can do to live more sustainably, including using energy-efficient light bulbs, designing your home eco-consciously, using solar power and saving water. These are all big issues that are often out of personal control on an everyday basis. To help address this, here are some quick, practical tips that can make a tangible difference on a personal level. Add indoor plants to freshen up your home naturally. Here are a few options:

- Mother-In-Law's Tongue: A perfect plant for a beginner, it can live in a variety of conditions. Be careful not to overwater it though.
- Peace Lily: It is quite easy to take care of. It clears the air really well and starts to droop when it needs water, making it easy for you to know when to nourish it. Make sure to wash your hands after touching this plant as it can cause minor irritation to some people.
- Chinese Evergreen: Another easy-to-maintain plant that loves low and medium-light environments. It cleans the air really well, grows a foot or two high, but just like the peace lily wash your hands if you feel any irritation.
- Spider Plant: Place this plant in a bright spot, but not in direct sunlight. It helps clear the air exceptionally well.
- Areca Palm: This plant is great if you think you may forget to water it occasionally. Keep in mind that it will grow better with a little water and a small amount of light. It will provide you with benefits that you may need if you spend a lot of time indoors.

Chapter 5: Gifting

1. Singh, Shilpi, 'How to Reduce Air Pollution and Waste Generation This Festive Season?' TERI (The Energy and Resources Institute), 23 October 2019, www.teriin.org/blog/how-reduce-air-pollution-and-waste-generation-festive-season.
2. Popescu, Alexandra, 'To Have and to Throw: Tackling Indian Wedding Food Waste', Food Tank, June 2019, foodtank.com/news/2019/06/to-have-and-to-throw-tackling-indian-wedding-food-waste/.
3. Discussed in Chapter 3.
4. Bratskeir, Kate, 'It's Officially Time to Ban Gift-Wrapping Paper', HuffPost Australia, 24 December 2018, www.huffingtonpost.com.au/entry/ban-gift-wrapping-paper_n_5c1d5c12e4b0407e907af36c?ri18n=true.
5. Blount, Sarah, ''Tis the Season . . . To Take Out the Trash?', NEEF, www.neefusa.org/holiday-waste.
6. 'Christmas Packaging Facts (2019): A Wide Range of Eye Opening Stats', GWP Group, 28 November 2019, www.gwp.co.uk/guides/christmas-packaging-facts/.
7. Kaushik, Divya. 'Make Your Wedding a Low-Waste Affair', *Times of India*, 11 January 2020, timesofindia.indiatimes.com/life-style/spotlight/make-your-wedding-a-low-waste-affair/articleshow/73187619.cms.
8. Ibid.
9. Pongrácz, Eva, The Environmental Impacts of Packaging, 2007, 10.1002/9780470168219.ch9.
10. Ibid.
11. Also, the global society can choose to make a broader difference in relation to life cycle analysis, discussed in Chapter 4 and broader discussions on societal changes discussed in Chapter 8 and Chapter 9.

12. A recycling centre's capacity will vary depending on where you are based in the world.

13. Clamshells are single-use packaging that is often clear on both sides. They comprise items such as scissors or other products where the manufacturer wants the consumer to be able to see the product within. They are made of single-use packaging that is recyclable but rarely is, as explained in the note below.

    See: Leahy, Stephen, 'This Common Plastic Packaging Is a Recycling Nightmare', National Geographic, 26 July 2019, www.nationalgeographic.com/environment/2019/07/story-of-plastic-common-clamshell-packaging-recycling-nightmare/.

14. Blister packs are usually a clear bubble of plastic mounted on a piece of cardboard. Batteries are often sold this way.

    See: Leahy, Stephen, 'This Common Plastic Packaging Is a Recycling Nightmare', National Geographic, 26 July 2019, www.nationalgeographic.com/environment/2019/07/story-of-plastic-common-clamshell-packaging-recycling-nightmare/.

15. 'There is a big demand for recycled PET (Polyethylene terephthalate) content in new clamshells, particularly those used in the food industry. However, clamshells aren't being recycled because they aren't being collected, and if collected, most existing material recovery facilities can't sort them out from the other materials. And if they can sort them, sticky labels or food residues can be a problem to remove, as is the cardboard backing on blister packs'.

    See: Leahy, Stephen, 'This Common Plastic Packaging Is a Recycling Nightmare', National Geographic, 26 July 2019, www.nationalgeographic.com/environment/2019/07/story-of-plastic-common-clamshell-packaging-recycling-nightmare/.

16. Leahy, 'This Common Plastic Packaging Is a Recycling Nightmare', National Geographic, www.nationalgeographic.com/environment/2019/07/story-of-plastic-common-clamshell-packaging-recycling-nightmare/.

17. The different types of plastics are discussed in the notes for Chapter 1.

18. Gibbens, Sarah, 'The Sticky Problem of Plastic Wrap', National Geographic, 12 July 2019, www.nationalgeographic.com/environment/2019/07/story-of-plastic-sticky-problem-of-plastic-wrap/.

19. 'There's a Science to Gift Giving: Experiences Are Better than Material Items', ScienceDaily, 15 December 2016, www.sciencedaily.com/releases/2016/12/161215143300.htm.

20. Findings from a 2018 survey conducted in the USA highlights a trend away from linear gifting practices to more sustainable options, especially experiences: 63 per cent of US adults (ages 18–65+) would prefer to receive an experience gift than a material gift (in the 2018) holiday season and 50 per cent of US adults plan to give experience gifts (in 2018). As many as 59 per cent of US adults believe that giving experience gifts is easier than shopping for material gifts. Over a third (36 per cent) of US adults want to

attend more events/live experiences in 2019. Almost 85 per cent of US adults agree that experience gifts are a great way for both the giver and receiver to enjoy something together. Also, 93 per cent of millennial women and 83 per cent of millennial men (ages 18 to 34) agree with that sentiment.

See: 'Tis the Season for Giving: New Survey Reveals Gifting Experiences on the Rise', Eventbrite, www.eventbrite.com/blog/press/press-releases/tis-season-giving-new-survey-reveals-gifting-experiences-rise/.

21. Rajappa, Amoolya, 'Raipur Hosts Country's First Ever Garbage Festival, Pulls off a 4-Day Fest That Celebrated the Best Out of Waste', YourStory, 24 January 2018, yourstory.com/2018/01/raipur-garbage-festival.

22. Mohan, Shruthi, 'How Bengaluru Soaked in Music and Recycled Art at Echoes of Earth, India's First and Only "Green Music" Fest', YourStory, 5 December 2018, yourstory.com/2018/12/echoes-of-earth-eco-friendly-music-festival.

23. Among other ideas, couples and wedding organizers are reducing waste through e-invites or seed paper invites, locally sourced flowers, minimum construction and reusable props for the decor. There are many ways to avoid plastic, too, especially for tables and wrapping. Renting and recycling outfits is also becoming a sustainable trend for weddings.

See: Kaushik, Divya, 'Make Your Wedding a Low-Waste Affair', *Times of India*, 11 January 2020, timesofindia.indiatimes.com/life-style/spotlight/make-your-wedding-a-low-waste-affair/articleshow/73187619.cms.

24. Seventeen zero-waste party rentals available in India are listed in the 'Zero-Waste Library' section at the end of this chapter.

25. S., Lekshmi Priya, 'Can a Zero-Waste Wedding Fit In Your Budget? These Folks Show You How!', Better India, 13 July 2018, www.thebetterindia.com/150443/news-zero-waste-wedding-chennai/.

26. Krishnan, Kripa, 'How Indian Couples Are Throwing Zero-Waste, Plastic-Free, Vegan Weddings', iDiva.com, 27 November 2019, www.idiva.com/weddings/inspiration/meet-the-indian-couples-opting-for-zero-waste-plastic-free-vegan-weddings/18004673.

27. Mishka, '7 Best Green Festivals in the World', 12 July 2011, Verdemode, www.verdemode.com/best-green-festivals-world/.

28. S., Lekshmi Priya, 'Can a Zero-Waste Wedding Fit In Your Budget? These Folks Show You How!', Better India, 13 July 2018, www.thebetterindia.com/150443/news-zero-waste-wedding-chennai/.

29. 'Fireworks: Their Impact on the Environment', Terrapass, 28 December 2016, www.terrapass.com/fireworks-impact-environment.

30. 'Diwali has created a persistent pollution problem in India for years. It's typically celebrated with firecrackers, which contribute to already massive amounts of smog in many Indian cities.'

See: Yeung, Jessie, 'Ahead of Diwali, New Delhi Residents Fear They'll Be Choking on Smog', CNN 26 October 2019, edition.cnn.com/2019/10/25/asia/india-diwali-pollution-intl-hnk-scli/index.html.

31. Further studies have been recommended to establish the exact extent of all factors.

32. Greven, Frans E., et al, 'Air Pollution During New Year's Fireworks and Daily Mortality in the Netherlands', Nature News, 5 April 2019, www.nature.com/articles/s41598-019-42080-6.

33. The air quality index measures the concentration of poisonous particulate matter in the air.

34. 'Diwali Fireworks Worsen Delhi's Hazardous Air Pollution', DW, www.dw.com/en/diwali-fireworks-worsen-delhis-hazardous-air-pollution/a-51013069.

35. Jain, Neha, 'Air Pollution Soars during Diwali in Rural Haryana,' Mongabay, 6 November 2018, india.mongabay.com/2018/11/air-pollution-soars-during-diwali-in-rural-haryana/.

36. Dutta, Saptarshi, 'Chennai Reels Under Air Pollution On Diwali, As Delhi And Mumbai Fare Better Than Last Year: News,' NDTV, 20 October 2017, swachhindia.ndtv.com/chennai-reels-under-air-pollution-on-diwali-as-delhi-and-mumbai-fare-better-than-last-year-13728/.

37. Cappucci, Matthew and Raloff, Janet, 'Holiday Fireworks Can Bring Extreme Pollution, India Finds', Science News for Students, 27 July 2019, www.sciencenewsforstudents.org/article/holiday-fireworks-can-bring-extreme-pollution-india-finds.

38. Ibid.

39. Roser, Max, 'Economic Growth', Our World in Data, 24 November 2013, ourworldindata.org/economic-growth.

40. Frost, Rosie, 'Waste from Worship: Solving India's Unique River Pollution Problem', Living, 26 February 2020, www.euronews.com/living/2020/02/26/waste-from-worship-solving-india-s-unique-river-pollution-problem.

41. Ibid.

42. This is discussed at length in Chapter 8.

43. A drone display was used in Shanghai, instead of fireworks, for New Year's Eve celebrations to mark the commencement of 2020. New technology, such as drones, provide potential alternatives to other products that create waste, while giving a parallel level of enjoyment.
See: Liberatore, Stacy, 'Shanghai Replaces Fireworks on New Year's Eve with Thousands for a Stunning Display,' Daily Mail Online, 2 January 2020, www.dailymail.co.uk/sciencetech/article-7845641/Shanghai-replaces-fireworks-New-Years-Eve-thousands-stunning-display.html.

44. Johnson, Bea, *Zero Waste Home: The Ultimate Guide to Simplifying Your Life*, Penguin, 2013.

45. 'Bamboo Speakers', Instructables, 8 October 2017, www.instructables.com/id/Bamboo-Speakers/.

46. 'Zero Waste Events-Cutlery Banks or Plate Rentals in India', Hungry Palette, 11 February 2019, www.thehungrypalette.com/blog/2018/12/29/zero-waste-events-cutlery-banks-or-plate-rentals-in-india.

47. S., Lekshmi Priya, 'Hello, Zero-Plastic Parties: 11 Urban Initiatives That Let You Rent Plates & Cutlery!', Better India, 27 September 2019, www.thebetterindia.com/182727/lifestyle-rent-cutlery-party-zero-plastic-party-sustainable-living-india/.

Chapter 6: Community

1. Sams, Lauren, 'Tackling India's e-Waste Recycling Crisis,' The University of Sydney, 27 March 2019, www.sydney.edu.au/news-opinion/news/2019/03/27/tackling-indias-ewaste-recycling-crisis.html.
2. Kumar, Suni, et al, 'Challenges and Opportunities Associated with Waste Management in India', Royal Society Open Science, March 2017, https://www.researchgate.net/publication/315541171_Challenges_and_opportunities_associated_with_waste_management_in_India.
3. 'The Ladies Compartment in Mumbai Local Trains', The Art Blog, Wovensouls, 7 February 2013, wovensouls.org/2013/02/07/the-ladies-compartment-mumbai-local-train/.
4. Dhar, Debanjan, '10 Interesting Facts about The Famous Dabbawallas of Mumbai', Storypick, 8 February 2016, www.storypick.com/dabbawalla-facts/.
5. '5 Management Lesson Form Mumbai Train', BrandTalk, 29 July 2019, brandtalk.co.in/blog/5-management-lesson-form-mumbai-train/.
6. Caldicott, Carolyn and Caldicott, Chris, 'Time for Tiffin: The History of India's Lunch in a Box', *Guardian*, 17 August 2014, www.theguardian.com/lifeandstyle/2014/aug/17/tiffin-the-history-of-indias-lunch-in-a-box-mumbai.
7. Thomke, Stefan, 'Mumbai's Models of Service Excellence', *Harvard Business Review*, 1 August 2014, hbr.org/2012/11/mumbais-models-of-service-excellence.
8. For more about SWaCH, visit swachcoop.com/.
9. For more about Saahas, visit saahas.org/.
10. 'Why We Do What We Do', Daily Dump, dailydump.org/about-us/.
11. 'The Most Common Types of Office Waste', Sydney City Rubbish, 6 February 2020, www.sydneycityrubbish.com.au/the-most-common-types-of-office-waste/.
12. 'Reducing Waste: What You Can Do', Environmental Protection Agency, 20 December 2019, www.epa.gov/recycle/reducing-waste-what-you-can-do.
13. Ibid.
14. Ibid.
15. 'Traffic-Related Air Pollution', Health Effects Institute, 31 July 2019, www.healtheffects.org/air-pollution/traffic-related-air-pollution.
16. These people are 44 per cent less likely to be overweight, 27 per cent less likely to have high blood pressure and 34 per cent less likely to have diabetes.

See: 'Taking Public Transportation Instead of Driving Linked with Better Health', ScienceDaily, 8 November 2015, www.sciencedaily.com/releases/2015/11/151108124754.htm.

17. 'Taking Public Transportation Instead of Driving Linked with Better Health', ScienceDaily, www.sciencedaily.com/releases/2015/11/151108124754.htm.

18. How Public Transit Helps Conserve Energy:
    - A bus with as few as seven passengers is more fuel-efficient than the average single-occupant automobile.
    - The fuel efficiency of a fully occupied bus is six times greater than that of the average single-occupant automobile.
    - The fuel efficiency of a fully-occupied train car is fifteen times greater than that of the average commuter's single-occupant automobile.
    - In terms of energy consumption per passenger mile (energy used to transport one passenger one mile), transit is more energy-efficient.
    - Buses use 8.7 per cent less energy per passenger mile than a typical automobile.
    - Commuter trains use 23.7 per cent less energy per passenger mile than a typical automobile.

19. How Public Transit Helps Reduce Air Pollution:
    - Public transit moves people efficiently while producing significantly less air pollution to move one passenger one mile—compared to moving a person one mile in a single-occupant automobile.
    - Buses emit only 20 per cent as much carbon monoxide per passenger mile as a single-occupant automobile.
    - Buses emit only 10 per cent as many hydrocarbons per passenger mile as a single-occupant automobile (hydrocarbons are VOCs—an ozone precursor).
    - Buses emit only 75 per cent as many nitrogen oxides (another ozone precursor) per passenger mile as a single-occupant automobile.
    - Trains emit only 25 per cent as many nitrogen oxides per passenger mile as a single-occupant automobile, and nearly 100 per cent less hydrocarbons and carbon monoxides.'

    See: DART First State, 'Division of Air and Waste Management: The Environmental Benefits of Riding Public Transit', State of Delaware, www.dnrec.delaware.gov/dwhs/info/Pages/OzonePublicTrans.aspx.

20. 'Environmental Benefits of Public Transportation', South University, 10 August 2016, www.southuniversity.edu/whoweare/newsroom/blog/environmental-benefits-of-public-transportation-31178.

21. Gaikwad, Vaibhav, 'Improving E-Waste Management in India', Australia India Institute, 16 January 2020, www.aii.unimelb.edu.au/publications/very-short-policy-brief/improving-e-waste-management-in-india/.

22. The launch of smartphones globally has created lines of hundreds, and even thousands, of consumers waiting to be one of the first ones to get their hands on a new phone.

See: 'Waiting For iPhone X: The One Line People Don't Seem To Mind', PYMNTS, 4 November 2017, www.pymnts.com/apple/2017/iphone-release-iphone-sales-news/.

23. Laprise, John, 'How Smartphones Are Changing the World', World Economic Forum, 5 December 2014, www.weforum.org/agenda/2014/12/how-smartphones-are-changing-the-world/.

24. Bhattacharjee, Yudhijit, 'Smartphones Revolutionize Our Lives, But At What Cost?', National Geographic, 25 January 2019, www.nationalgeographic.com/science/2019/01/smartphones-revolutionize-our-lives-but-at-what-cost/.

25. 'UN Environment Chief Warns of "Tsunami" of e-Waste'", United Nations, 5 May 2015, www.un.org/sustainabledevelopment/blog/2015/05/un-environment-chief-warns-of-tsunami-of-e-waste-at-conference-on-chemical-treaties/.

26. Ibid.

27. Ibid.

28. 'Only 20 per cent of global e-waste is formally recycled. The remaining 80 per cent is often incinerated or dumped in landfills. Many thousands of tonnes also find their way around the world to be pulled apart by hand or burned by the world's poorest workers. This crude form of urban mining has consequences for people's well-being and creates untold pollution.'
See: Ryder, Guy and Houlin, Houlin Zhao, 'The World's e-Waste Is a Huge Problem. It's Also a Golden Opportunity', World Economic Forum, 24 January 2019, www.weforum.org/agenda/2019/01/how-a-circular-approach-can-turn-e-waste-into-a-golden-opportunity/.

29. Two organizations focusing on improving infrastructure to do with waste are Hasiru Dala Innovations and Plastics for Change.
See: 'Waste Management Company in Bangalore: Total Waste Management Solution', hasirudalainnovations.com/ and,
'Reduce Plastic Pollution by Recovery & Recycling', Plastics for Change, www.plasticsforchange.org/.

30. Ryder, Guy and Houlin, Houlin Zhao, 'The World's e-Waste Is a Huge Problem. It's Also a Golden Opportunity', World Economic Forum, 24 January 2019, www.weforum.org/agenda/2019/01/how-a-circular-approach-can-turn-e-waste-into-a-golden-opportunity/.

31. 'Electronic Waste', World Health Organization, 18 October 2019, www.who.int/ceh/risks/ewaste/en/.

32. Ibid.

33. Broader scale impacts are discussed in Chapters 7 and 8.

34. Gomes, Elton, 'This Microfactory Is Turning e-Waste into Reusable Material', World Economic Forum, 12 April 2018, www.weforum.org/agenda/2018/04/can-e-waste-be-converted-into-reusable-material-this-indian-origin-scientist-just-launched-the-world-s-first-microfactory-to-do-just-that.

35. The 2016 figures for the nine countries producing the maximum e-waste (in million metric tonnes) are:
   1) China: 7.211.
   2) USA: 6.295.
   3) Japan: 2.139.
   4) India: 1.975.
   5) Germany: 1.884.
   6) UK: 1.632.
   7) Brazil: 1.534.
   8) Russian Federation: 1.392.
   9) France: 1.373.
   See: Gomes, Elton, 'This Microfactory Is Turning e-Waste into Reusable Material', World Economic Forum, 12 April 2018, www.weforum.org/agenda/2018/04/can-e-waste-be-converted-into-reusable-material-this-indian-origin-scientist-just-launched-the-world-s-first-microfactory-to-do-just-that.

36. The US Bureau of Labour Statistics tracks prices for broad categories of goods over time. For the eighteen years up to 2015, prices have dropped dramatically in almost every tech sector. Notably, the drop in computer hardware is particularly steep.
   See: Rosoff, Matt, 'Why Is Tech Getting Cheaper?', World Economic Forum, 16 October 2015, www.weforum.org/agenda/2015/10/why-is-tech-getting-cheaper/.

37. Gomes, Elton, 'This Microfactory Is Turning e-Waste into Reusable Material', World Economic Forum, 12 April 2018, www.weforum.org/agenda/2018/04/can-e-waste-be-converted-into-reusable-material-this-indian-origin-scientist-just-launched-the-world-s-first-microfactory-to-do-just-that.

38. A key example of focusing on creating structure for waste workers, women and girls in particular, is WEIGO (Women in Informal Employment: Globalizing and Organizing), a global research and policy network focusing on improving livelihoods of the working poor in the informal economy. In Pune, this has led to collaboration with the '8000-strong union of waste pickers, Kagad Kach Patra Kashtakari Panchayat (KKPKP) and its solid waste management cooperative SWaCH' to improvements of occupational health and safety standards, and provided support for KKPKP's campaign on extended producer responsibility, which directly improves the livelihoods of women and girls. Such collaborations are the starting point to ensure that waste workers, especially the most vulnerable, are included in a system that values each segment (the circular economy) rather than dumping waste down the socioeconomic scale (the linear process) where the most vulnerable sections of Indian society, and other communities globally, are impacted the most. Simple steps with strong support from a variety of stakeholders present an opportunity to

implement a more sustainable and equal environment for every single man, woman, boy and girl.
See: Alfers, Laura, 'OHS in India', *WIEGO*, www.wiego.org/ohs-india.

39. A broader change is discussed in depth in Chapters 7 and 8.

40. Leighton, Mara, '16 Ways You're Wasting Resources and Money Every Day without Realizing It', Business Insider, 26 June 2018, www.businessinsider.com/everyday-ways-to-reduce-waste-2017?r=AU&IR=T#1-straws-1.

41. Broom, Douglas, 'This Indian School Accepts Plastic Waste Instead of Fees', World Economic Forum, 29 May 2019, www.weforum.org/agenda/2019/05/this-indian-school-accepts-plastic-waste-instead-of-fees/.

42. Think Change India, 'This RJ-Turned Businessman's Juice Corner in Bengaluru Is Promoting Zero-Waste and Sustainability', YourStory, 22 July 2019, yourstory.com/socialstory/2019/07/juice-corner-bengaluru-zero-waste-eat-raja.

43. Keetley, Amanda, '9 Ways to Reduce Plastic in Your Workplace,' Less Plastic, 27 July 2018, www.lessplastic.org.uk/9-ways-to-reduce-plastic-in-your-workplace/.

44. Ibid.

45. Ibid.

46. One area not discussed while talking about habits that lead to pollution is smoking: 'Smokers around the world buy roughly 6.5 trillion cigarettes each year . . . Cigarette filters are made of a plastic called cellulose acetate. When tossed into the environment, they dump not only that plastic, but also the nicotine, heavy metals, and many other chemicals they've absorbed into the surrounding environment.'

The story behind the plastic in cigarettes is this: Plastic filters were invented in the 1950s in response to lung cancer fears. By the mid-1960s, researchers realized that the substances being filtered, like nicotine, were what made cigarettes satisfying, so manufacturers made filters less effective. Today, 98 per cent of cigarette filters are made of plastic fibres. Approximately 4.5 trillion cigarette butts are discarded each year worldwide, making them the most littered item on earth. If collected, they are not often recycled, while the toxins from the product can leach harmful chemicals, which last up to a decade, into waterways.'
See: Root, Tik, 'Cigarette Butts Are Toxic Plastic Pollution. Should They Be Banned?', National Geographic, 9 August 2019, www.nationalgeographic.com/environment/2019/08/cigarettes-story-of-plastic/.

47. For more, visit hasirudalainnovations.com/about-us/.

48. For more, visit www.skrap.in/.

49. Ibid.

50. For more, visit y-east.org/about.

51. For more, visit anthillcreations.org/.

52. Govind, Ranjani, 'Go Green with Garden', *The Hindu*, 14 May 2019, www.thehindu.com/life-and-style/vani-murthys-waste-segregation-and-terrace-garden/article27127457.ece.

53. R, Sasha, '"Compostwali" Poonam Bir Kasturi of Daily Dump Tells You How to Lead a More Environmentally Conscious Life', YourStory, 7 June 2019, yourstory.com/herstory/2019/06/compostwali-poonam-bir-kasturi-daily-dump-environment.

54. For more, visit hasirudalainnovations.com/about-us/.

55. 'Manik Thapar, Founder, Eco Wise Waste Management Pvt. Ltd', YourStory, 29 March 2010, yourstory.com/2010/03/manik-thapar-founder-eco-wise-waste-management-pvt-ltd.

56. Karelia, Gopi and Bhaskar, Sonia, 'A Comprehensive Waste Management Model Needed to Cut Down Plastic Generation: Dia Mirza', NDTV, 20 April 2018, swachhindia.ndtv.com/comprehensive-waste-management-model-needed-cut-plastic-generation-dia-mirza-13141/.

57. For more, visit www.cseindia.org/page/board-members.

58. *Hidden Kingdom* by Nirupa Rao, Paper Planes, www.joinpaperplanes.com/shop/books-and-magazines/art/hidden-kingdom-nirupa-rao/.

59. Mankani, Sneha and Manghnani, Devika, 'Vogue Warriors: Meet the Film Producer Who Wants to Put Food on Every Plate in Mumbai', *Vogue India*, 4 May 2020, www.vogue.in/culture-and-living/content/vogue-warriors-pragya-kapoor-film-producer-ek-saath-foundation-covid-19-relief-mumbai.

60. For more, visit www.apollo.io/people/Nayantara/Jain/55c828fb73696410dc26b402.

61. 'Wilma Rodrigues', The Real Deal, NDTV, 3 April 2016, sites.ndtv.com/therealdeal/contestants/wilma-rodrigues/.

62. For more, visit www.cbramkumar.com/.

63. For more, visit www.ournativevillage.com/about-us/about-us.html.

64. For more, visit www.binbag.in/.

65. For more, visit karosambhav.com/.

66. For more, visit namoewaste.com/index.php/about/mission/.

67. 'GasLand', IMDb, 17 January 2011, www.imdb.com/title/tt1558250/.

68. 'Dark Waters', IMDb, 27 November 2019, www.imdb.com/title/tt9071322/?ref_=fn_al_tt_1.

69. 'The True Cost', IMDb, 29 May 2015, www.imdb.com/title/tt3162938/?ref_=fn_al_tt_1.

70. 'An Inconvenient Truth', IMDb, 30 June 2006, www.imdb.com/title/tt0497116/?ref_=fn_al_tt_1.

71. 'Tomorrow', IMDb, 2 December 2015, www.imdb.com/title/tt4449576/?ref_=fn_al_tt_2.

72. '2040', IMDb, 23 May 2019, www.imdb.com/title/tt7150512/?ref_=fn_al_tt_1.

73. For more, visit https://aggie-horticulture.tamu.edu/Kindergarden/CHILD/COM/COMMUN.HTM

Chapter 8: City

1. Mingaleva, Zhanna; Vukovic, Natalya; Volkova, Irina and Salimova, Tatiana, 'Waste Management in Green and Smart Cities: A Case Study of Russia', *Sustainability*, Vol. 12, Issue 1, 2019, https://www.mdpi.com/2071-1050/12/1/94.

2. Ahuja, Aastha, 'Fighting Mountains of Garbage: Here Is How Indian Cities Dealt With Landfill Crisis In 2018', NDTV, 3 January 2019, swachhindia. ndtv.com/year-ender-2018-waste-management-landfill-how-indian-cities-dealt-with-landfill-crisis-29247/.

3. Mohan, Shruthi, 'Even after Human Chain Steels Show, Karnataka Government Firm on Building Steal Flyover', YourStory, 20 October 2016, yourstory.com/2016/10/human-chain-govt-firm-on-steal-flyover.

4. 'Experts Reject Idea of Steel Flyover, Elevated Corridor, Say Congestion Will Increase in Bengaluru', *New Indian Express*, 21 January 2019, wgbis.ces. iisc.ernet.in/energy/wetlandnews/News-January2019/Jan%2021%202019/ Experts%20reject%20idea%20of%20steel%20flyover%20The%20New%20 Indian%20Express%20Jan%2021%202019.pdf.

5. Safi, Michael, 'Mumbai Beach Goes from Dump to Turtle Hatchery in Two Years', *Guardian*, 30 March 2018, www.theguardian.com/world/2018/ mar/30/mumbai-beach-goes-from-dump-to-turtle-hatchery-in-two-years.

6. Dalton, Jane, 'Sea Turtles Return to Nest on Beach Which Used to Be a Dumping Ground for Plastic', *Independent*, 30 March 2018, www. independent.co.uk/news/world/asia/turtles-hatchlings-born-mumbai-india-versova-beach-litter-cleanup-a8281266.html.

7. 'Citizen Matters Contest: Master the Master Plan for a Chance to Win 10K', Citizen Matters, 21 February 2020, bengaluru.citizenmatters.in/ master-the-master-plan-contest.

8. 'Introducing the 2016 Goldman Prize Winners', Goldman Environmental Foundation, 17 August 2016, www.goldmanprize.org/blog/introducing-the-2016-goldman-prize-winners/.

9. For more, visit www.evlogiaeco.com/.

10. For more, visit www.kabadiwallaconnect.in/.

11. For more, visit www.papermanfoundation.org/about/.

12. For more, visit streemuktisanghatana.org/.

13. Sattiraju, Nikitha, 'What Makes These 10 Beautiful Cities the Most Sustainable in the World?', YourStory, 8 September 2016, yourstory. com/2016/09/10-most-sustainable-cities.

14. Dalkmann, Holger, and Prabhu, Ashwin, '5 Keys to Sustainable Development in Indian Cities', World Resources Institute, 22 April 2013, www.wri.org/ blog/2013/04/5-keys-sustainable-development-indian-cities.

15. '20 Minute Neighbourhoods', Plan Melbourne, www.planmelbourne.vic.gov. au/current-projects/20-minute-neighbourhoods.

16. Other areas of focus that will allow a move towards more sustainable practices within a city, such as segregating waste, limiting the amount of non-recyclable material used and managing resources like new technology (e- waste, for instance), which in turn assist people, have been discussed in earlier chapters. You can refer to them to help your city achieve more sustainable practices.

17. Sustainable travel is also discussed in other chapters.

18. First discussed in Chapter 4.

19. Sussmann, Adrian, 'Why Mobility in Urban Environments Is Crucial to the Evolution of Cities', Smart Cities World, 29 May 2019, www.smartcitiesworld.net/opinions/opinions/why-mobility-in-urban-environments-is-crucial-to-the-evolution-of-cities.

20. Moavenzadeh, John and Corwin, Scott, '10 Ways Government Leaders Can Improve Transport Mobility', World Economic Forum, 30 March 2018, www.weforum.org/agenda/2018/03/transport-commute-systems-government-simsystem.

21. Both domestic and international migration has been a source of growth in urban areas, and has brought both opportunities and challenges to cities, migrants and governments. Municipal authorities are needed to manage the growing trends and will rely on accurate data on migration and urbanization. Poignantly, this data is not often available, leading to an inability to plan and implement sound systems that could improve liveability. This area needs to improve for sustainable cities to develop worldwide.

    See: 'Urbanization and Migration', Migration Data Portal, migrationdataportal.org/themen/urbanisierung-und-migration.

22. Statistically, 30 per cent of the world population lived in urban areas in the 1950s, 55 per cent in 2018. Estimates suggest that by 2030, 60 per cent of the world population will live in urban settings.

    See: 'Urbanization and Migration', Migration Data Portal, migrationdataportal.org/themen/urbanisierung-und-migration.

23. By 2030, ten more cities are expected to gain megacity status (the criteria is to have in excess of 10 million people), including Hyderabad and Ahmedabad in India, Chengdu and Nanjing in China, Seoul in South Korea, Ho Chi Minh City in Vietnam and Tehran in Iran. Other cities will move dramatically on the list due to a higher rate of population growth, including Manila in the Philippines and Jakarta in Indonesia, while both Karachi and Lahore in Pakistan will continue to see population increases.

    See: Thornton, Alex, '10 Cities Are Predicted to Gain Megacity Status by 2030', World Economic Forum, 6 February 2019, www.weforum.org/agenda/2019/02/10-cities-are-predicted-to-gain-megacity-status-by-2030.

24. Note that as economies grow, allowing people to earn more money, their desire will change. Data-backed research has highlighted that there are four main economic levels that countries transition through based on the money that a person generates as income per day: people on Level 1 have less than

US $2, people on Level 2 have between US $2 and US $8, people on Level 3 have between US $8 and US $32, while the top echelon, Level 4, which includes only one billion people (compared to over 6 billion people on other levels), have in excess of US $32. Notably, only half a century ago over half of the world population was on Level 1. In 2010, that figure has dropped to 13 per cent. When figures per country are analysed, it is an average. For example, people can live in a Level 2 country with a Level 4 income.

Poignantly, when people move from one level to another, their aspirations change, such as owning motor vehicles. A large issue concerning overall traffic congestion is that the more people have their incomes increased, the more vehicles will be on the road, causing more congestion, air pollution and waste. This area needs to be addressed by systemic changes towards different methods of mobility and functions within a city in order for people to live sustainably.

See: Rosling, Hans, et al, *Factfulness: Ten Reasons We're Wrong about the World-and Why Things Are Better than You Think*, Sceptre, 2018.

25.   Wolf, Harrison, 'How Aerial Transportation Will Shape Cities of the Future', World Economic Forum, 13 November 2019, www.weforum.org/agenda/2019/11/flying-taxis-drones-cities-right-rules/.

26.   The following list comprises statistics on the eight most congested cities worldwide in the following format: Name of the city: hours lost in congestion: last mile travel speed.

1)   Moscow, Russia: 210: 11.

2)   Istanbul, Turkey: 157: 10.

3)   Bogotá, Colombia: 272: 7.

4)   Mexico City, Mexico: 218: 9.

5)   Sao Paulo, Brazil: 154: 10.

6)   London, England: 227: 7.

7)   Rio De Janeiro: 199: 13.

8)   Boston, US: 164: 11.

See: Wolf, Harrison, 'How Aerial Transportation Will Shape Cities of the Future', World Economic Forum, 13 November 2019, www.weforum.org/agenda/2019/11/flying-taxis-drones-cities-right-rules/.

27.   Wolf, 'How Aerial Transportation Will Shape Cities of the Future', World Economic Forum, www.weforum.org/agenda/2019/11/flying-taxis-drones-cities-right-rules/.

28.   'Division of Air and Waste Management,' The Environmental Benefits of Riding Public Transit, www.dnrec.delaware.gov/dwhs/info/Pages/OzonePublicTrans.aspx.

29.   Tyres consist of 19 per cent natural rubber and 24 per cent synthetic rubber, a plastic polymer. The remaining is made up of metal and other compounds.

30.   Root, Tik, 'Tires: The Plastic Polluter You Never Thought About,' National Geographic, 20 September 2019, www.nationalgeographic.com/environment/2019/09/tires-unseen-plastic-polluter/

31. 'Designing a Seamless Integrated Mobility System (SIMSystem): A Manifesto for Transforming Passenger and Goods Mobility', January 2018, www3.weforum.org/docs/Designing_SIMSystem_Manifesto_ Transforming_Passenger_Goods_Mobility.pdf.

32. Maria Gallucci, 'An Increasingly Urbanized Latin America Turns to Electric Buses', Yale Environment 360, 9 September 2019, e360.yale.edu/features/ an-increasingly-urbanized-latin-america-turns-to-electric-buses

33. Teh, David and Khan, Tehmina, 'As Cities Grow, the Internet of Things Can Help Us Get on Top of the Waste Crisis', Conversation, 31 January 2020, theconversation.com/as-cities-grow-the-internet-of-things-can-help-us- get-on-top-of-the-waste-crisis-127917.

34. Increased levels of heat are produced due to large areas taken over by concrete buildings, bitumen, etc. that absorb heat instead of reflecting it, like forests do. The heat island contributes to climate change, which in turn create problems such as the floods witnessed in India in 2019 and the bush fires in Australia and Brazil in the same year.

35. Jim Morrison, 'Can We Turn Down the Temperature on Urban Heat Islands?', Yale Environment 360, 12 September 2019, e360.yale.edu/features/ can-we-turn-down-the-temperature-on-urban-heat-islands

36. Vaishnavi Chandrashekhar, 'As the Monsoon and Climate Shift, India Faces Worsening Floods', Yale Environment 360, 17 September 2019, e360.yale.edu/ features/as-the-monsoon-and-climate-shift-india-faces-worsening-floods

37. Huang, Chyi-Yun and Cantada, Isabel, 'Challenges to Implementing Urban Master Plans–What Are We Missing?', World Bank Blogs, 25 February 2019, blogs.worldbank.org/sustainablecities/challenges-implementing- urban-master-plans-what-are-we-missing.

38. Research has highlighted, from an in-depth look at Tanzania (the example is being used to illustrate a template for other developing countries around the world), that some critical reasons for the ineffectiveness of master plans implemented in developing cities are:
   1) An inherent weaknesses in the master plans themselves,
   2) A disconnect between spatial planning, sector or infrastructure plans and budgeting and investment planning decisions,
   3) A lack of coordination among key agencies,
   4) A lack of, or ineffective development controls,
   5) Unrealistic planning standards and regulations,
   6) Limited capacity and resources for enforcement.
   Solutions and approaches to planning need to be innovative but also practical. Also, more can be done to empower local authorities as they have the keenest on-the-ground knowledge.
   See: Huang, Chyi-Yun and Cantada, Isabel, 'Challenges to Implementing Urban Master Plans–What Are We Missing?', World Bank Blogs, 25 February 2019, blogs.worldbank.org/sustainablecities/challenges- implementing-urban-master-plans-what-are-we-missing.

39. Mead, Nick Van, '22 Of World's 30 Most Polluted Cities Are in India, Greenpeace Says', *Guardian*, 5 March 2019, www.theguardian.com/ cities/2019/mar/05/india-home-to-22-of-worlds-30-most-polluted-cities-greenpeace-says

40. 'India Population (LIVE)', Worldometers, www.worldometers.info/world-population/india-population/.

41. Ranges of waste pollution, a per capita average, spread from 0.65 kg in Sub-Saharan Africa to 0.95 kg in East Asia and the Pacific to 1.1 kg in locations such as Central and Western Asia, Latin America and the Middle East. The OECD (countries throughout Europe, North America, Australia and New Zealand, Chile, Japan and South Korea.) countries generate 2.2 kg waste per capita per day.
See: 'Urban Development Series Knowledge Papers, Chapter 3: Waste Generation', siteresources.worldbank.org/INTURBANDEVELOPMENT/ Resources/336387-1334852610766/Chap3.pdf.

42. 'Urban Development Series Knowledge Papers, Chapter 3: Waste Generation', siteresources.worldbank.org/INTURBANDEVELOPMENT/ Resources/336387-1334852610766/Chap3.pdf.

43. Ibid.

44. Sharholy, Mufeed; Ahmad, Kafeel; Mahmood, Gauhar and Trivedi, R., 'Municipal Solid Waste Management in Indian Cities: A Review', https:// www.researchgate.net/publication/6394075_Municipal_solid_waste_ management_in_Indian_cities_-_A_review

45. 'Our Planet Is Drowning in Plastic Pollution', www.unenvironment.org/ interactive/beat-plastic-pollution/.

46. Ibid.

47. For more, visit https://plasticoceans.org/the-facts/.

48. Dunham, Will, 'World's Oceans Clogged by Millions of Tons of Plastic Trash', Scientific American, 12 February 2015, www.scientificamerican.com/ article/world-s-oceans-clogged-by-millions-of-tons-of-plastic-trash/.

49. Rosling, Hans, et al, *Factfulness: Ten Reasons We're Wrong about the World–and Why Things Are Better than You Think*, Sceptre, 2018.

50. Noted as a practical view of 'hope' that is approached with accurate and up-to-date information rather than a fairy tale-type 'everything will be alright' scenario.

51. Rosling, Hans, et al, *Factfulness: Ten Reasons We're Wrong about the World–and Why Things Are Better than You Think*, Sceptre, 2018.

52. Dandapani, Swetha, 'Unpaid and Undervalued, How India's Waste Pickers Fight Apathy to Keep Our Cities Clean', News Minute, 30 November 2017, www.thenewsminute.com/article/oppressed-and-unrecognised-life-waste-pickers-crucial-india-s-sanitation-72426

53. Reddy, Akhileshwari, 'A Law for Waste Pickers', *Down To Earth*, 12 April 2018, www.downtoearth.org.in/news/waste/a-law-for-waste-pickers-60103.

54. 'India generates around 150,000 tonnes of solid waste each day—the equivalent of 15,000 trucks carrying 10 tonnes each. And the problem is growing as the country's population swells: one in every six people on the planet live in India, a population that has doubled in size over the past forty years. By 2050, the World Bank estimates India's waste could be three-and-a-half times greater than today, which is indicative of a wider global trend. Rapid population growth and urban migration is expected to increase annual waste generation to 3.4 billion tonnes by 2050, a 70 per cent increase on 2016 levels. These predictions assume business as usual. But if more cities follow Indore's lead (discussed in the main text) and change how their waste is managed, it could help contain the problem.' See: Wood, Johnny, 'These Are India's Cleanest Cities', World Economic Forum, 4 October 2019, www. weforum.org/agenda/2019/10/india-clean-cities/.

55. Kelly, Jeremy, 'These Are the World's Most Future-Proof Cities', World Economic Forum, 11 June 2018, www.weforum.org/agenda/2018/06/ worlds-most-future-proof-cities-jll/.

56. Fitzgerald, Sunny, '25 Places That Have Committed to Going Zero-Waste', 29 October 2018, www.nationalgeographic.com/travel/lists/zero-waste-eliminate-sustainable-travel-destination-plastic/.

57. Wood, Johnny, 'These Are India's Cleanest Cities', World Economic Forum, 4 October 2019, www.weforum.org/agenda/2019/10/india-clean-cities/.

58. Ibid.

59. Davies, Nick and Hardman, Micheal, 'Why More Cities Should Adopt Green Roofs', 29 October 2019, www.weforum.org/agenda/2019/10/urban-green-roofs-trees-environment/.

60. Byington, Cara Cannon, 'Time to Bust the Silos: Coral Reefs, Human Health + Sewage Pollution', Cool Green Science, 26 April 2019, blog. nature.org/science/science-brief/time-to-bust-the-silos-coral-reefs-human-health-sewage-pollution/?utm_source=cgs&utm_medium=archive&utm_campaign=Big per cent2BQuestions.

61. Globally, there are numerous recommendations to break down the silo effect of addressing problems. This situation sees individual stakeholders or groups of stakeholders attempting to address issues without linking up with or assisting other parties who are trying to do similar things. Additionally, this case is exacerbated when organizations or governments, for instance, are tackling different symptoms of the same cause. Recommendations made by global bodies, such as the United Nations, who have implemented frameworks such as the Sustainable Development Goals have urged parties to collaborate more in order to limit this silo effect.
See: Byington, Cara Cannon, 'Time to Bust the Silos: Coral Reefs, Human Health + Sewage Pollution', Cool Green Science, 26 April 2019, blog. nature.org/science/science-brief/time-to-bust-the-silos-coral-reefs-human-health-sewage-pollution/?utm_source=cgs&utm_medium=archive&utm_campaign=Big per cent2BQuestions.

62.   'Water Pollution', WWF, wwf.panda.org/knowledge_hub/teacher_resources/webfieldtrips/water_pollution/.

63.   The waste problem is not restricted to waterways. On land, and in the air, waste is prevalent due to current systems. 'Land pollution [is] broadly classified as municipal solid waste, construction and demolition waste or debris, and hazardous waste. Municipal solid waste includes non-hazardous garbage, rubbish and trash from homes, institutions (e.g., schools), commercial establishments and industrial facilities.' In locations where the population is a lot, in Delhi in India, for example, the scale of land-based waste can truly be seen in the Ghazipur trash dump which is reportedly just months away (as of June 2019) from rising higher than the Taj Mahal.' This is a profound example of a system that does not pay attention to the detrimental effects of a linear approach. Understanding the extent of land pollution, no matter what category it falls under, is vital to addressing the situation. From a purely financial perspective, across the globe the 'value of ecosystems to human livelihoods and well-being is 125 trillion USD per year.'
        See: Nathanson, Jerry A., 'Land Pollution', Encyclopædia Britannica, www.britannica.com/science/land-pollution,
        'Garbage Mountain at Delhi's Ghazipur Landfill to Rise Higher than Taj Mahal by 2020,' *Hindustan Times*, 4 June 2019, www.hindustantimes.com/india-news/garbage-mountain-at-delhi-s-ghazipur-landfill-to-rise-higher-than-taj-mahal-by-2020/story-RC0kwZdUmdHHfDs3rJGngI.html and,
        'Goal 15: Life on Land', UNDP, www.undp.org/content/undp/en/home/sustainable-development-goals/goal-15-life-on-land.html.

64.   'River Pollution: Causes, Actions and Revival', Janhit Foundation, www.janhitfoundation.in/pdf/booklet/river_pollution_causes_action_and_revival.pdf.

65.   'Trash Collecting Boats Are Cleaning India's Rivers', Futurism, YouTube, 21 November 2017, www.youtube.com/watch?v=AzijVxKvI18.

66.   'Cradle to Cradle as a Toolbox for the Circular Economy', EPEA, 6 July 2017, epea-hamburg.com/circular-economy/.

67.   Searra, Imogen, 'From Avocado Pits to Sustainable Straws', *Getaway*, 7 February 2019, www.getaway.co.za/travel-news/from-avocado-pits-to-sustainable-straws/.

68.   For more, visit www.biofase.com.mx/copia-de-home-2.

69.   Gray, Alex, 'This Plastic Bag Is 100 Per Cent Biodegradable', World Economic Forum, 4 May 2018, www.weforum.org/agenda/2018/05/this-plastic-bag-is-100-biodegradable-and-made-of-plants/.

70.   Similarly, larger and more efficient turbines for wind energy have increased the demand for forms of technologies that do not produce waste like fossil fuels do. These demands are predominantly generated by people looking to reduce the effects of wasteful practices, many of which lead to environmental issues such as climate change. The use of renewable energy is a key way to

reduce the heat island effect in urban centres, in combination with making cities green in various ways.

See: Wood, Johnny, 'India Is Now Producing the World's Cheapest Solar Power', World Economic Forum, 28 June 2019, www.weforum.org/agenda/2019/06/india-is-now-producing-the-world-s-cheapest-solar-power?utm_source=Facebook%2BVideos&utm_medium=Facebook%2BVideos&utm_campaign=Facebook%2BVideo%2BBlogs.

71. In 2017, the greenest cities named at the global C40 summit (the group consists of over ninety cities that brings together over 650 million citizens aiming to reduce climate change) are:

- Copenhagen, Denmark
- Chicago, USA
- Dar es Salaam, Tanzania
- New York, USA
- Auckland, New Zealand.

All of these cities, as well as others, aspiring for greener practices, approach methods that suit their situation while being aware of people and the planet. See: Pardo, Diana, 'Which Are the 10 Greenest Cities in the World?', Smart City Lab, 16 August 2019, www.smartcitylab.com/blog/urban-environment/these-are-the-10-greenest-cities-in-the-world/.

72. Menon, Anil, 'Cities Can Become Smarter, by Going Circular,' World Economic Forum, 21 March 2018, www.weforum.org/agenda/2018/03/how-to-make-cities-smarter-circular/.

73. Ibid.

74. 'Circular Cities', Ellen MacArthur Foundation, www.ellenmacarthurfoundation.org/our-work/activities/circular-economy-in-cities.

75. For more, visit www.theuglyindian.com/.

76. For more, visit www.unitedwaymumbai.org/cleanshores.

77. For more, visit indiaenvironment.org/.

78. For more, visit shuddhi.ngo/.

79. For more, visit www.aahwahan.com/.

80. For more, visit, www.earthlingsngo.com/.

81. 'Clean Mumbai, Happy Mumbai: Here's How You Can Be a Part of The Mahim Beach Clean Up', LBB, Mumbai, lbb.in/mumbai/mahim-beach-clean-up/.

82. 'Tips for Organizing a Beach Clean-Up', NEEF, www.neefusa.org/nature/land/tips-organizing-beach-clean.

83. 'How to Create a Campaign: Step-by-Step Guide', Southern Regional Education Board, 10 December 2015, www.sreb.org/how-create-campaign-step-step-guide.

84. Shah, Sneha, 'List of Leading Waste Management Companies in India–Interesting Ways to Re-Use Ways: Green World Investor', Green World Investor, 6 June 2018, www.greenworldinvestor.com/2018/06/06/list-of-

leading-waste-management-companies-in-india-interesting-ways-to-re-use-ways/.

Chapter 9: Travel

1. 'Solid Waste Management', UN Environment Programme, www.unenvironment.org/explore-topics/resource-efficiency/what-we-do/cities/solid-waste-management.

2. 'Solid Approach to Waste: How 5 Cities Are Beating Pollution', UN Environment Programme, www.unenvironment.org/news-and-stories/story/solid-approach-waste-how-5-cities-are-beating-pollution.

3. Chitra, 'Karnataka Style Holige/Obbattu/Puran Poli Recipe', Chitra's Food Book, 1 September 2019, www.chitrasfoodbook.com/2015/03/holigeobbattupuran-poli-recipe-ugadi.html.

4. Latha, 'Bele Holige Recipe: How to Make Bele Obbattu: Tasty Puran Poli', 21 March 2017, vegrecipesofkarnataka.com/262-bele-holige-recipe-bele-obbattu-pooran-poli.php.

5. For more, visit app.zerowastehome.com/.

6. 'It's Official—Spending Time Outside Is Good for You', ScienceDaily, 6 July 2018, www.sciencedaily.com/releases/2018/07/180706102842.htm.

7. Raising of livelihood levels have been found to reduce large environmental issues that are often associated with poor countries, such as mismanaged waste. Notably, individuals in high-income levels produce large amounts of waste, but they are also in a position to make change as opposed to subsistence farmers, for example.

8. 'What a Waste: An Updated Look into the Future of Solid Waste Management', World Bank, 20 September 2018, www.worldbank.org/en/news/immersive-story/2018/09/20/what-a-waste-an-updated-look-into-the-future-of-solid-waste-management.

9. 'Solid Waste Management', UN Environment Programme, www.unenvironment.org/explore-topics/resource-efficiency/what-we-do/cities/solid-waste-management.

10. Hellyer, Isabelle, 'Fashion Creates More Greenhouse Pollution Than the Airline Industry', Vice, 5 December 2017, www.vice.com/en_asia/article/bjdx95/fashion-creates-more-greenhouse-pollution-than-the-airline-industry.

11. Diop, Sarah-Aby and Shaw, Peter J., 'End of Use Textiles: Gifting and Giving in Relation to Societal and Situational Factors', *Detritus*, 31 March 2018, digital.detritusjournal.com/articles/end-of-use-textiles-gifting-and-giving-in-relation-to-societal-and-situational-factors/13.

12. Single-use plastic waste has discussed in detail in the earlier chapters.

13. Marine debris is waste that ends up in oceans and other large water bodies.

14. 'Great Pacific Garbage Patch', National Geographic Society, 9 October 2012, www.nationalgeographic.org/encyclopedia/great-pacific-garbage-patch/.

15. Ibid.
16. Ibid.
17. Ibid.
18. The four classified categories of plastic within the patch are:
    - Microplastics (0.05- 0.5 cm)
    - Mesoplastics (0.5–5 cm)
    - Macroplastics (5–50 cm), and
    - Megaplastics (above 50 cm).

    See: 'The Great Pacific Garbage Patch,' The Ocean Cleanup, theoceancleanup. com/great-pacific-garbage-patch/.
19. 'About 54 per cent of the debris . . . comes from land-based activities in North America and Asia. The remaining . . . comes from boaters, offshore oil rigs, and large cargo ships that dump or lose debris directly into the water. The majority of this debris—about 705,000 tons—is fishing nets.'

    See: 'Great Pacific Garbage Patch', National Geographic Society, 9 October 2012, www.nationalgeographic.org/encyclopedia/great-pacific-garbage-patch/.
20. 'Great Pacific Garbage Patch', National Geographic Society, 9 October 2012, www.nationalgeographic.org/encyclopedia/great-pacific-garbage-patch/.
21. The following are a number of unusual items found due to ships losing cargo: 'In 1990, five shipping containers of Nike sneakers and work boots were lost . . . In 1992, rubber duckies floated in the Pacific when a ship lost tens of thousands of bathtub toys. The ducks were accompanied by turtles, beavers, and frogs . . . In 1994, a ship lost 34,000 pieces of hockey gear, including gloves, chest protectors, and shin guards'.

    See: 'Great Pacific Garbage Patch', National Geographic Society, 9 October 2012, www.nationalgeographic.org/encyclopedia/great-pacific-garbage-patch/.
22. 'The Great Pacific Garbage Patch is not the only marine (waste) vortex, it's just the biggest. The Atlantic and Indian Oceans both have (waste) vortexes. Even shipping routes in smaller bodies of water, such as the North Sea, are developing garbage patches.' See: 'Great Pacific Garbage Patch', National Geographic Society, 9 October 2012, www.nationalgeographic.org/encyclopedia/great-pacific-garbage-patch/.
23. Edmond, Charlotte, 'This Is What the World's Waste Does to People in Poorer Countries', World Economic Forum, 16 May 2019, www.weforum.org/agenda/2019/05/this-is-what-the-world-s-waste-does-to-people-in-poorer-countries/.
24. Ferronato, Navarro and Torretta, Vincenzo, 'Waste Mismanagement in Developing Countries: A Review of Global Issues', *International Journal of Environmental Research and Public Health*, 24 March 2019, www.ncbi.nlm.nih.gov/pmc/articles/PMC6466021/.
25. The company promotes long-term use of products, which includes understanding the value of the item and repairing it when it becomes damaged rather than throwing it away.

See: 'Worn Wear: Better Than New', Patagonia Outdoor Clothing & Gear, www.patagonia.com/worn-wear.html.

26.   Here are a few examples:

a.   British start-up Winnow has developed smart meters that analyse our waste. They are used in commercial kitchens to measure what food gets thrown away and then identify ways to reduce waste.

b.   Dutch company DyeCoo has developed a process of dyeing cloth that uses no water and no chemicals other than the dyes themselves.

c.   Close The Loop, an Australian company, has spent more than a decade recovering value from old printer cartridges and soft plastics. Their new innovation turns these materials into roads. The products are mixed in with asphalt and recycled glass to produce a higher-quality road surface that lasts up to 65 per cent longer than traditional asphalt.

d.   Canadian firm Enerkem has turned the concept of making waste run a car into reality. Their technology extracts the carbon from waste that can't be recycled. It then takes five minutes to turn the carbon into a gas that can be used to make biofuels like methanol and ethanol, as well as chemicals which can be used in thousands of everyday products.

e.   Schneider Electric uses recycled content and recyclable materials in its products, prolongs product lifespan through leasing and pay-per-use, and has introduced take-back schemes into its supply chain.

f.   Cambrian Innovation's EcoVolt technology treats wastewater contaminated by industrial processes, not just turning it into clean water, but even producing biogas that can be used to generate clean energy.

g.   Lehigh Technologies turns old tyres and other rubber waste into something called micronized rubber powder, which can then be used in a wide variety of applications from tyres to plastics, asphalt and construction material.

h.   HYLA Mobile works with many of the world's leading manufacturers and service providers to repurpose and reuse either the devices themselves, or their components. It's estimated that more than 50 million devices have been reused, making US $4 billion for their owners and stopping 6500 tonnes of e-waste from ending up in landfill.

i.   TriCiclos began in Chile in 2009 with the stated aim of working towards a 'world without waste'. Since then it has built and operated the largest network of recycling stations in South America, diverting 33,000 metric tonnes of recyclable material from landfill and saving over 140,000 metric tonnes of carbon emissions.

j.   Miniwiz has created the Trashpresso machine which is the ultimate expression of sustainable upcycling. It is a mobile upcycling plant that can be transported in two shipping containers to its customers. Once there, it turns 50 kg of plastic bottles an hour into a low-cost building material, using no water and only solar power.

    k.   AB in Bev, the world's largest brewer, wants 100 per cent of its product to be in packaging that's returnable or made from majority-recycled content by 2025. Already nearly half of its drinks are sold in returnable glass bottles.

27.   For more, visit theoceancleanup.com/about/.

28.   For updates, visit theoceancleanup.com/updates/.

29.   'Landfills: An Unsustainable Form of Waste Management', Hazardous Waste Experts, 11 March 2014, www.hazardouswasteexperts.com/landfills-an-unsustainable-form-of-waste-management/.

30.   'Scientists attribute the global warming trend observed since the mid-twentieth century to the human expansion of the "greenhouse effect" warming that results when the atmosphere traps heat radiating from Earth towards space. Certain gases in the atmosphere block heat from escaping. Long-lived gases that remain semi-permanently in the atmosphere and do not respond physically or chemically to changes in temperature are described as "forcing" climate change . . . human activities are changing the natural greenhouse. Over the last century, the burning of fossil fuels like coal and oil has increased the concentration of atmospheric carbon dioxide (CO_2). This happens because the coal or oil burning process combines carbon with oxygen in the air to make CO_2. To a lesser extent, the clearing of land for agriculture, industry, and other human activities has increased concentrations of greenhouse gases.'.
See: 'The Causes of Climate Change', NASA, 30 September 2019, climate.nasa.gov/causes/.

31.   This has been discussed in detail in earlier chapters.

32.   Holbole, Ramesh, 'In Drought-Prone India, a Hard-Pressed Village Struggles to Survive', Yale E 360, 1 August 2019, e360.yale.edu/features/in-drought-prone-india-a-hard-pressed-village-struggles-to-survive

33.   Babu, Suresh, 'Farmer Suicides: A Call to Climate Action for India', ReliefWeb, 6 September 2017, reliefweb.int/report/india/farmer-suicides-call-climate-action-india.

34.   'The Waste Hierarchy', NSW Environment Protection Authority, 21 September 2017, www.epa.nsw.gov.au/your-environment/recycling-and-reuse/warr-strategy/the-waste-hierarchy.

35.   'Localhood is a long-term vision that supports the inclusive co-creation of our future destination. A future destination where human relations are the focal point. Where locals and visitors not only co-exist, but interact around shared experiences of localhood. Where our global competitiveness is underpinned by our very own localhood. And where tourism growth is co-created responsibly across industries and geographies, between new and existing stakeholders, with localhood as our shared identity and common starting point'. For more, visit localhood.wonderfulcopenhagen.dk/.

36.   World Travel & Tourism Council, 'Five Reasons Sustainability Is Essential for Travel & Tourism', Medium, 29 March 2019, medium.com/@WTTC/five-reasons-sustainability-is-essential-for-travel-tourism-1dbd20f327bf.

37. Schwab, Klaus, 'Why We Need the "Davos Manifesto" for a Better Kind of Capitalism', World Economic Forum, 1 December 2019, www.weforum.org/agenda/2019/12/why-we-need-davos-manifesto-for-better-kind-of-capitalism/.

38. Ibid.

39. The shipping industry, unlike the technology available for aviation, is one area that could make a significant shift towards renewable technology, thus reducing waste from fossil fuels. Added to this, if they practice standards discussed in the Davos Manifesto, they would seek more accountability for overall actions. A detailed brief recommends 'to policy makers on promoting realistic renewable energy solutions, which can support energy efficiency and reduced emissions in the sector' including:

    • 'Renewable power applications for ships of all sizes include options for primary or hybrid propulsion, as well as on-board and shore-side energy use.

    • Renewables can be integrated through retrofits to the existing fleet or incorporation into new shipbuilding and design, with a small number of new ships striving for 100 per cent renewable energy or zero-emissions technology for primary propulsion in the long run.

    • Certain applications offer immediate growth potential, even if the contribution of renewables to the energy mix of the shipping sector remains limited in the near and medium terms.'

    See: 'Renewable Energy Options for Shipping', IRENA (International Renewable Energy Agency), February 2015, www.irena.org/publications/2015/Feb/Renewable-Energy-Options-for-Shipping.

40. Ferronato, Navarro, Torretta, Vincenzo Torretta, 'Waste Mismanagement in Developing Countries: A Review of Global Issues', International Journal of Environmental Research and Public Health, 24 March 2019, www.ncbi.nlm.nih.gov/pmc/articles/PMC6466021/.

41. A 2019 National Geographic survey of 3500 adults in the USA revealed strong support for sustainability. That's the good news. The challenge will be helping travellers take meaningful actions. According to the survey, while 42 per cent of US travellers would be willing to prioritize sustainable travel in the future, only 15 per cent of these travellers are sufficiently familiar with what sustainable travel actually means.

42. Fleming, Sean, 'People Can Swap Plastic Waste for Rice in This Philippines Community', World Economic Forum, 16 September 2019, www.weforum.org/agenda/2019/09/in-this-philippines-community-people-can-swap-plastic-waste-for-rice.

43. The airport has banned all single-use plastic items, including disposable cutlery made of thermocol (polystyrene or plastic), PET/PETE bottles (less than 200 ml), plastic bags (with/without handle), disposable dish/bowl for food packaging, straws and bubble wraps.

44. 'Mumbai Airport to Go Plastic-Free from Tomorrow, 1st Time Offenders to Be Fined Rs 5000', News18, 1 October 2019, www.news18.com/news/india/

mumbai-airport-to-go-plastic-free-from-tomorrow-1st-time-offenders-to-be-fined-rs-5000-2329311.html.

45. Slotnick, Davidm 'Plastic Water Bottles Are Banned at San Francisco Airport Starting This Week - Here's What You Need to Know', Business Insider, 21 August 2019, www.businessinsider.com/plastic-water-bottle-airport-ban-san-francisco-sfo-2019-8.

46. Matamua, Taito-Taaalii and Lionel. 'Renewing Materials: 3D Printing and Distributed Recycling Disrupting Samoa's Plastic Waste Stream', Victoria University of Wellington, 2015, researcharchive.vuw.ac.nz/handle/10063/5013.

47. Borunda, Alejandra, 'See How Much of the Amazon Is Burning, How It Compares to Other Years', National Geographic, 29 August 2019, www.nationalgeographic.com/environment/2019/08/amazon-fires-cause-deforestation-graphic-map/.

48. 'Australia Fires: A Visual Guide to the Bushfire Crisis', BBC, 31 January 2020, www.bbc.com/news/world-australia-50951043.

49. Tabuchi, Hiroko, 'Oil Giants, Under Fire From Climate Activists and Investors, Mount a Defense', New York Times, 23 September 2019, www.nytimes.com/2019/09/23/climate/oil-industry-climate-investment.html.

50. Christensen, Jen, 'The Amount of Plastic in the Ocean Is a Lot Worse than We Thought', CNN, 16 April 2019, edition.cnn.com/2019/04/16/health/ocean-plastic-study-scn/index.html.

51. Schwab, Klaus, 'Why We Need the "Davos Manifesto" for a Better Kind of Capitalism', World Economic Forum, 1 December 2019, www.weforum.org/agenda/2019/12/why-we-need-davos-manifesto-for-better-kind-of-capitalism/.

52. Numerous voices have been heard about the issues relating to waste globally. In 2019, Sir David Attenborough noted, 'You can do more and more and more the longer you live, but the best motto to think about is not waste things. Don't waste electricity, don't waste paper, don't waste food. Live the way you want to live, but just don't waste. Look after the natural world, and the animals in it, and the plants in it too. This is their planet as well as ours. Don't waste them.'
See: Abbott, Kate, '"Just Don't Waste": David Attenborough's Heartfelt Message to Next Generation', Guardian, 18 October 2019, www.theguardian.com/tv-and-radio/2019/oct/19/just-dont-waste-david-attenborough-advice-bbc-seven-worlds-one-planet.

53. '10 x 10 CHALLENGE', Style Bee, 30 October 2019, www.stylebee.ca/10-x-10-challenge/.

54. For more, visit indiahikes.com/.

55. Bhattacharyya, Smita, 'Arunachal Ride to Light Homes, Lure Tourists', Telegraph, 24 August 2018, www.telegraphindia.com/states/north-east/arunachal-ride-to-light-homes-lure-tourists/cid/1397544.

56. For more, visit www.goheritagerun.com/.

57. For more, visit agreenventure.in/.
58. For more, visit www.plasticsforchange.org/.
59. For more, visit repurpose.global/.
60. For more, visit www.plasticpollutioncoalition.org/about-us.
61. For more, visit www.plasticsoupfoundation.org/en/about-us/.
62. For more, visit www.gacircular.com/.

Further Reading: Recommended Resources

1. Shiva, Vandana, *Soil Not Oil*, North Atlantic Books (Penguin Random House), www.penguinrandomhouse.com/books/535624/soil-not-oil-by-vandana-shiva/.

2. Guha Ramachandra, *Environmentalism: A Global History*, ramachandraguha.in/archives/books/environmentalism-a-global-history.

3. Ramesh Mridula, *The Climate Solution: India's Climate-Change Crisis and What We Can Do About It*, Goodreads, www.goodreads.com/book/show/40101150-the-climate-solution.

4. Guha Ramachandra, *The Unquiet Woods: Ecological Change and Peasant Resistance in the Himalaya*, Goodreads, www.goodreads.com/book/show/116707.The_Unquiet_Woods.

5. Carson, Rachel, *Silent Spring by Rachel Carson*, Goodreads, www.goodreads.com/book/show/27333.Silent_Spring?from_search=true&from_srp=true&qid=xBYgQS5V0N&rank=1.

6. McDonough, William, *Cradle to Cradle: Remaking the Way We Make Things*, Goodreads, www.goodreads.com/book/show/5571.Cradle_to_Cradle.

7. Louv, Richard, *Last Child in the Woods*, richardlouv.com/books/last-child/.

8. McKibben, Bill, *Deep Economy: The Wealth of Communities and the Durable Future*, billmckibben.com/deep-economy.html.

9. Klein, Naomi, *This Changes Everything*, thischangeseverything.org/book/.

10. Thunberg, Greta, *No One Is Too Small to Make a Difference*, Penguin Books Australia, www.penguin.com.au/books/no-one-is-too-small-to-make-a-difference-9780141992716.

11. Friedman, Thomas L., *Hot, Flat, and Crowded: Why We Need a Green Revolution – and How It Can Renew America*, Goodreads, www.goodreads.com/book/show/2358737.Hot_Flat_and_Crowded.

12. Auyero, Javier, *Flammable: Environmental Suffering in an Argentine Shantytown*, Goodreads, www.goodreads.com/book/show/5627444-flammable.

13. For more, visit www.trashonomics.in/.

14. For more, visit upcyclerslab.com/.

15. *Hidden Kingdom* by Nirupa Rao, Paper Planes, www.joinpaperplanes.com/shop/books-and-magazines/art/hidden-kingdom-nirupa-rao/.

16. Sangameswaran, Shubhashree, *Let's Talk Trash: An Illustrated Handbook with Ideas towards a Less Messy World*, Goodreads, www.goodreads.com/book/show/41086098-let-s-talk-trash.